S0-BOR-164

INTRODUCTION TO COMPUTATIONAL METHODS FOR STUDENTS OF CALCULUS

INTRODUCTION TO COMPUTATIONAL METHODS FOR STUDENTS OF CALCULUS

SAMUEL S. McNEARY

Drexel University

PRENTICE-HALL, INC.

Englewood Cliffs, New Jersey

Library of Congress Cataloging in Publication Data

McNeary, Samuel S

Introduction to computational methods for students of calculus.

Bibliography: p.

1. Electronic data processing—Calculus.
2. Electronic data processing—Numerical analysis.
I. Title. II. Title: Computational methods for students of calculus.

QA303.M2 515'.028'54 72-771

ISBN 0-13-479501-6

10 9 8 7 6 5 4 3 2 1

Printed in the United States of America

PRENTICE-HALL INTERNATIONAL, INC., *London*
PRENTICE-HALL OF AUSTRALIA, PTY. LTD., *Sydney*
PRENTICE-HALL OF CANADA, LTD., *Toronto*
PRENTICE-HALL OF INDIA PRIVATE LIMITED, *New Delhi*
PRENTICE-HALL OF JAPAN, INC., *Tokyo*

CONTENTS

PREFACE

The advent of digital computers, readily accessible to students, has created an interest in using these facilities to support and enhance undergraduate courses in science and engineering.

This book is written primarily for a course in computational methods that is given parallel to the freshman calculus course. The interplay between these two courses should not only result in a development of computational skills but provide an added depth of understanding of basic concepts of analysis. For those who have passed the freshman calculus stage, this text could serve as an introduction to a study of numerical analysis. Except for ordinary differential equations, the coverage of this text is compatible with Course B4 of the A.C.M. Curriculum 68.

Techniques of computation and computations relevant to elementary analysis are the emphasis of this book. In form it presents a panorama of computational algorithms and supporting lists of problems. It is not intended as a text in programming. Rudiments of the language Fortran are presented here in outline form with illustrative examples in programming situations relevant to the solution of exercises in the remainder of the text. It is also not intended as a text in numerical analysis, but may be considered as a useful preface to one. The ACM Curriculum 68, Course B4 plays this preparatory role for a course in numerical analysis.

The author wishes to acknowledge the helpful suggestions of the late Professor George E. Forsythe of Stanford University and Professor Albert Herr of Drexel University.

Samuel S. McNeary

1

INTRODUCTION

Today's student in engineering and science requires an expanded view of computational problems and solution techniques as a prelude to his undergraduate studies. This student has had or is having a first course in calculus, and has been required in such a course, or in other courses, to perform simple computations. In his subsequent undergraduate work, particularly in engineering, he will find an increasing need to produce computed problem solutions efficiently. He will need experience in the use of computing machinery.

No prior experience in programming a digital computer has been assumed. There are presented here only a few definitions and examples of the language Fortran, so that some of the simplest computations can be performed as soon as possible. We will build our skills with this language as we proceed to widen our range of problem solving. It is not the intention of this book to make programming the major objective. Enough will be learned of Fortran to meet modest needs. The primary objective is to build a wide computational experience that has a direct bearing on other undergraduate courses of study. Reference will be made to several excellent textbooks on programming that will enable anyone interested to become expert.

Fortran has been selected from the many computer languages because of its wide usage in scientific computation. The language that an individual will use depends first on whether he has access to a computing machine through an "on-line" typewriter input/output or through input prepared "off line," for example punched cards. Of course the language selected depends also on what language can be translated by the available machine, and on the user's preference and job requirements.

If resonable facility is obtained with one language, then another is easily learned. Our major problem will be to construct correct and efficient programs for the solution of the problems in this book.

The digital computer is our primary laboratory instrument, but we should not rule out or underrate pencil, paper, and slide rule. Indeed these aids are sufficient for many computational problems. It is evident folly to spend an hour of our own time and the resources of a multimillion-dollar computational facility to solve a problem that needs only a moment with a slide rule. Aside from wasteful usage of human and machine time, harm is sometimes done by naively assuming that all the digits that the machine prints are meaningful to our problem because they are produced by an impressive piece of equipment.

The computer has indeed opened an unexpected Pandora's box, in that some sources of error, not ordinarily troublesome in hand computation, have become amplified and are of major concern in machine computation. For example, the subtraction of two nearly equal numbers reveals a substantial increase in relative error. The subsequent use of this difference propagates its error through the computation. In hand computations such a situation is apparent and ad hoc measures can sometimes be taken to avoid this loss of significant figures. In a machine computation such a situation can easily go undetected, and the solver merely has in hand a print out of an obviously wrong result, with no hint as to the cause of the error.

A great asset offered by a computing machine—arithmetic operations at very low cost and at high speed—also brings with it a liability. Numbers in a machine are stored with a fixed number of significant digits, and arithmetic operations on these numbers introduces error. For example, a machine having a two-digit register would perform the multiplication problem given as follows:

$$(.35*10^{-3})(.75*10^{2}) = .25*10^{-1}$$

The correct result, however, is $.2625*10^{-1}$. If a very large number of arithmetic operations are performed, the error may grow so that at some point all subsequent numbers may have no significant digits.

Although computation is an ancient human skill, courses like this one are new on the scene. There are obvious reasons: the current economic and scientific demands, the general availability of the modern digital computer, and certainly the inherent academic merit of these courses. The computer is not necessary to the presentation of a good calculus course, but for most students it has been found to be an asset. Added insight into numerous concepts of classical analysis is available to the student through finite computations with a finite set of numbers. The concepts of a convergent sequence of numbers, of the derivative function, of the mean value of the derivative of a function over an interval, of a function expressed in terms of an integral—all can be vitalized by computed approximations for specific examples.

Newton's name is attached to several important computational procedures in this book. This is partial evidence that the algorithms that we will use have been known for a long time. However, in contrast to our predecessors of even 25 years ago, we can use these algorithms to explore and experiment in interesting computational problems.

Some of the algorithms and problems in this book can also be found in textbooks on numerical analysis. In this book, however, we generally limit our study to a stated algorithm and selected problems that conform to the conditions stated in the algorithm. Very little is said about proofs for convergent processes unless they are consistent with, and would enhance, a basic calculus course. Little is said about the problem of error analysis or control, and nothing about numerical processes that are inherently unstable. All these topics are the substance of a course in numerical analysis.

For a basic course in computation it is appropriate to include numerical integration. If course hours are limited and deletions prove necessary, it is best to start with Chapter 3 (Fortran) and proceed to sample the subsequent lists of computational opportunities, so that integration may be included.

There is a large volume of current literature in computation. Some is understandable and meaningful to the undergraduate, but for the most part it is written by and for professionals with specific technical interests and wide mathematical background. At the end of the book there is a list of references selected to be of interest to those who are beginning a computer-assisted study of science and engineering.

What constitutes a solvable problem? How do we assess the availability and validity of input data? What are appropriate objectives? What are valid and feasible procedures? How do we assess the validity of a proposed computed solution? Do we have the personnel and machine facilities to perform a proposed computation? Is the objective worth the effort and expense? Is it best to first solve a simpler problem to verify our procedures and program? These large questions are obviously not answered here, but after some computational experience, we may be in a better position to at least consider the merit of asking such questions, and seeking answers to them.

2

TYPES OF COMPUTATIONS

It would be difficult and without much merit to attempt the assignment of every computational problem to a specific category, but some distinctions are worth consideration.

Computations may be direct or iterative. We shall call a *direct* computation one based on an algorithm (a rule, or a set of directions) containing a finite number of arithmetic steps. An example is the computation of your weighted grade-point average.

An *iterative* computation is a sequence of direct computations. The elements of the sequence hopefully converge to the desired result. An example of an iterative computation is the algorithm learned in grade school for the approximation of a square root. For $\sqrt{2}$, this procedure establishes the sequence of rational numbers 1.0, 1.4, 1.41, 1.414, 1.4142, ...

2.1 Examples of Direct Computation of an Exact Solution

2.1.1 Find the roots of $2x^2 + 7x - 15 = 0$.

$$2[x^2 + \tfrac{7}{2}x - \tfrac{15}{2}] = 0$$

$$2[x^2 + \tfrac{7}{2}x + \tfrac{49}{16} - \tfrac{169}{16}] = 0$$

$$2[(x + \tfrac{7}{4})^2 - (\tfrac{13}{4})^2] = 0$$
$$2(x + 5)(x - \tfrac{3}{2}) = 0$$

Thus if $2x^2 + 7x - 15 = 0$, then $x = -5$ or $x = \frac{3}{2}$.

Here we have performed a specified, finite number of arithmetic operations on the ordered set $(2, 7, -15)$ to determine the numbers -5 and $\frac{3}{2}$.

2.1.2 Find the number pair (x, y) that satisfies the linear conditions

$$2x - 3y = 11$$
$$x + 2y = -5$$

Cramer's rule is one of several algorithms that leads to the equivalent system

$$x = \begin{vmatrix} 11 & -3 \\ -5 & 2 \end{vmatrix} \Big/ \begin{vmatrix} 2 & -3 \\ 1 & 2 \end{vmatrix} = \frac{(11)(2) - (-3)(-5)}{(2)(2) - (-3)(1)} = \frac{7}{7} = 1$$
$$y = \begin{vmatrix} 1 & 11 \\ 1 & -5 \end{vmatrix} \Big/ \begin{vmatrix} 2 & -3 \\ 1 & 2 \end{vmatrix} = \frac{(2)(-5) - (11)(1)}{7} = \frac{-21}{7} = -3$$

This algorithm involves a finite number of arithmetic operations on the array of numbers

$$\begin{bmatrix} 2 & -3 & 11 \\ 1 & 1 & -5 \end{bmatrix}$$

to exactly determine the ordered pair $(1, -3)$.

2.2 Examples of Direct Computation of a Specified Approximation

2.2.1 If $f(x) = 1/(1 + x)$ find the approximate value of $f'(\frac{1}{2})$ for $\Delta x = \frac{1}{10}$.

$$f'(\tfrac{1}{2}) = \frac{f(\frac{1}{2} + \frac{1}{10}) - f(\frac{1}{2})}{\frac{1}{10}} + \text{error}$$
$$= 10\left(\frac{1}{1 + \frac{6}{10}} - \frac{1}{1 + \frac{1}{2}}\right) + \text{error}$$
$$= -\tfrac{5}{12} + \text{error}$$

Does the value of "error" depend upon the arithmetic that has been done? From the limit $f'(x)$ determined in calculus, what is the value of "error"?

Another example is the approximation of the number $\int_a^b f(x)\,dx$ by the partitioning of the interval $[a, b]$ into n equal subdivisions, each of length Δx, and with midpoints $x_1, x_2, \ldots, x_n$:

$$\int_a^b f(x)\,dx \approx \sum_{i=1}^{n} f(x_i)\,\Delta x = \Delta x \sum_{i=1}^{n} f(x_i)$$

2.2.2 Find an approximate value of $\int_0^1 (1/(1+x))dx$, where $n = 2$.

The area of the figure bounded by the hyperbola $f(x) = 1/(1+x)$ and $y = 0$, $x = 0$, $x = 1$ is defined as the number

$$\int_0^1 (1/(1+x))\,dx.$$

The value of this definite integral may be approximated by the sum of the areas of the rectangles indicated in Fig. 2.1.

x_i	$f(x_i)$
$\frac{1}{4}$	$\frac{1}{1+\frac{1}{4}} = \frac{4}{5}$
$\frac{3}{4}$	$\frac{1}{1+\frac{3}{4}} = \frac{4}{7}$

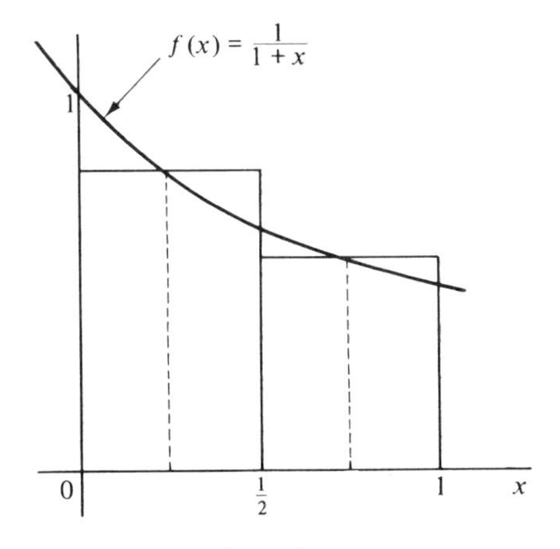

Fig. 2.1

$$\sum_{i=1}^{2} f(x_i) = \frac{48}{35}$$

$$\int_0^1 \frac{dx}{1+x} = \frac{1}{2} \cdot \frac{48}{35} + \text{error}$$

$$= \frac{24}{35} + \text{error}$$

Does the value of "error" depend on the arithmetic that has been done? With a table of logarithms can you estimate the value of "error"?

2.3 An Iterative Computation with a Prescribed Number of Steps

2.3.1 Assume that for the sequence $\{a_n\}$,

$$a_1 = 2, \quad a_{n+1} = \tfrac{1}{2}\left(a_n + \frac{10}{a_n}\right), \qquad n = 1, 2, 3, \ldots$$

then $\lim_{n\to\infty} a_n = \sqrt{10}$. Find a_4.

n	a_n	$\frac{10}{a_n}$	$a_n + \frac{10}{a_n}$
1	2	5	7
2	$\frac{7}{2}$	$\frac{20}{7}$	$\frac{89}{14}$
3	$\frac{89}{28}$	$\frac{280}{89}$	$\frac{15{,}761}{2{,}492}$
4	$\frac{15{,}761}{4{,}984}$		

For most purposes, it is preferable to perform this computation with the aid of a slide rule. This method approximates each of these rational numbers by another rational number, a decimal having three significant figures.

n	a_n	$\frac{10}{a_n}$	$a_n + \frac{10}{a_n}$
1	2.	5.	7.
2	3.50	2.86	6.36
3	3.18	3.15	6.33
4	3.16	3.16	

Let the error be defined as $E = a_n - \sqrt{10}$. In the first table, if we use the arithmetic of rationals, E depends only on n. In the second computation, E depends not only on n but on the rule for rounding off (approximating) these rational numbers at each step.

2.4 An Iterative Computation Terminated by a Prescribed Upper Bound for Absolute Error

2.4.1 For the sequence defined in problem 2.3.1, find a_n if $|E_n| < .05$.

After finding a_n at each iteration, we must be able to determine E_n sufficiently well to ascertain whether or not the condition $|E_n| < .05$ is satisfied. For this particular example, the error E can be approximately determined as follows:

$$E_n = a_n - \sqrt{10}$$
$$E_n(a_n + \sqrt{10}) = a_n^2 - 10$$
$$E_n(a_n + a_n - E_n) = a_n^2 - 10$$
$$E_n \simeq \frac{a_n^2 - 10}{2a_n}, \qquad \text{if } E_n \text{ is small}$$

If we use the slide rule,

n	a_n	$\frac{10}{a_n}$	$a_n + \frac{10}{a_n}$	a_n^2	$a_n^2 - 10$	$2a_n$	E_n	
1	2	5	7					
2	3.5			12.2	2.2	7.0	0.314	$\lvert E_2 \rvert \nless 0.05$
		2.86	6.36					
3	3.18			10.1	0.1	6.36	0.016	$\lvert E_3 \rvert < 0.05$

Thus, the solution to this problem is $a_3 = 3.18$, since $E_3 = .016$ satisfies the condition $|E_n| < .05$.

2.5 Problems

2.5.1 Find the approximation of $\int_0^2 (1/(1 + x))\, dx$ using the rule of problem 2.2.2 and $n = 4$. Repeat this calculation using a slide rule.

2.5.2 Find, by slide rule, $\sqrt{55}$ with $|E_n| < .05$ using the iteration

$$a_1 = 5, \quad a_{n+1} = \tfrac{1}{2}\left(a_n + \frac{55}{a_n}\right), \qquad n = 1, 2, 3, \ldots$$

2.5.3

(a) Draw the graphs of $x - 2y + 2 = 0$, $2x + 6y - 21 = 0$.
(b) Find the solution by a direct computation.
(c) The sequence $\{x_n, y_n\}$, for which

$$x_1 = 4, \quad y_n = \frac{7}{2} - \frac{x_n}{3}, \quad x_{n+1} = 2y_n - 2, \qquad n = 1, 2, 3, \ldots$$

is such that the solution found in (b) is $\lim_{n\to\infty} (x_n, y_n)$. Find (x_4, y_4).

2.5.4

(a) Draw the graphs of $x^2 = y + 1$, $y^2 = x + 2$.
(b) Try to find the solution in the first quadrant by a direct computation involving the elimination of one unknown.
(c) The sequence $\{x_n, y_n\}$, for which

$$x_1 = 1, \quad y_1 = 1, \quad x_{n+1} = \sqrt{y_n + 1},$$
$$y_{n+1} = \sqrt{x_{n+1} + 2}, \qquad n = 1, 2, 3, \ldots$$

is such that the solution is $\lim_{n\to\infty} (x_n, y_n)$. Find (x_5, y_5) using a slide rule.

2.5.5 Using instructions similar to those of problem 2.5.4, find, by slide rule, an approximate solution of

$$x = 1 - \sqrt{y}$$
$$y = \sqrt{4 - x^2}$$

3

FORTRAN

3.1 The Fortran Language

The computer has become the essential laboratory instrument for students of science and engineering. It is not only a device that accepts and stores data but also a set of instructions for a computational procedure. The earliest of the electronic digital computers could only perform a fixed sequence of arithmetic operations on given data. The essential capabilities of a modern computer appeared about 1949, when machines were built that would not only retain data and a set of computational instructions in storage but that would also modify these instructions as directed by intermediate computed results. In virtually all our problem solutions we will use this ability of the computer to change the order of executing instructions and to appropriately alter instructions as the solution proceeds.

Each computer has a *machine language*, which is its basic medium of communication with its user. The manufacturer of the machine has a primary concern for the structure of this language in relationship to the design of the equipment.

To the novice, a machine language seems unduly detailed and tedious to use. The first attempt to construct *compilers* or *translators* was made in

1953. To extend the usage of computers to the common man, languages were developed using some of the familiar symbolism and syntax of English and algebra. Programs were then written that would translate a set of instructions written in one of these problem-oriented languages into an equivalent set of instructions expressed in the machine language. An important aspect of the current computer industry is the development of *software*, that is, the programming devices to facilitate the usage of computer *hardware*.

Today there are many simplified languages translated into many machine languages. Some are FORTRAN (FORmula TRANslator), PL/1 (Programming Language 1), COBOL (COmmon Business Oriented Language), ALGOL (ALGorithmic Oriented Language), BASIC, and APL (A Programming Language). The latter two are primarily adapted to the restrictions of the typewriter as an input/output device. In our discussion we will use Fortran, a currently popular language oriented toward scientific use.

Fortran is a language that has a vocabulary of few words and a syntax of simple structure but that requires precision of its users. An academic by-product of learning to solve problems with a computer is the technique of using language precisely when addressing the computer. This technique contrasts with our usual lack of precision in conversing with another person, where we depend on his ability to ascribe meaning to our words. In addition, we learn the necessity of presenting to the computer a logically consistent sequence of commands for the machine to execute.

For effective use of a computer facility, certain initial problems must be surmounted. They are listed here not only to forewarn but also to allay fears. To start from scratch and arrive at a state in which one can consider the computer as his useful laboratory ally is not unduly difficult and is full of surprising rewards.

1. A necessary minimum of vocabulary and syntax of the programming language must be learned.

2. For the problem to be solved a correct (but not necessarily efficient) program must be constructed.

The remainder of this chapter presents a useful subset of the language Fortran along with examples of programs for simply stated types of problems that occur frequently. The remainder of the book contains computational problems for which logically correct and, as experience grows, efficient programs are to be constructed and given the ultimate test—solution by the computer.

3. The procedures of the computer facility must be known.

4. Enough of the operation of the key-punch machine or typewriter input to meet the user's needs must be known.

The usual practice in a computation center open to student use is to supply printed rules and procedures for the facility and to maintain a bulletin board to keep its public up to date. Similarly, if the user of the facility is to keypunch his own cards or use a typewriter input, printed instructions are usually available. Of these tasks, the most formidable—keypunching cards—is simple indeed. If printed instructions are not at hand, sufficient competence can be gained by 10 minutes' worth of watching an experienced operator and a little practice on your own.

5. Both the novice and the experienced user must be prepared to accept disappointment, in that first efforts are not always rewarded by success.

There is always the possibility of an error in keypunching or in program logic that escapes even the most diligent proofreader. Some problems that appear on first examination to be without computational complications may turn out to involve excessive error or even some aspect of noncomputability.

The sober reflection of things gone wrong and a revision of the program with the incorporation of corrective measures is called *debugging*, an activity of *all* who compute.

The Fortran compiler is helpful in that when a program is rejected for errors in vocabulary and syntax and for certain logical errors, it states its reasons for rejection. A version of Fortran called Watfor or Watfiv, available on some larger machines, provides excellent editing assistance. On many occasions this makes debugging a simple task.

To enable you to solve problems on the computer as soon as possible, there is in this chapter an outline of some of the words and grammar of Fortran and some illustrative programs. There is much more to be learned of this language and its application, and, as need arises, we may make appropriate additions to our knowledge of Fortran. For those who wish to become fluent or who need reference material, there is a list of relevant publications at the end of the book.

As an illustration of the coinage used in our transactions with the machine, there follows a reproduction of the punched-card input for a familiar problem and the printed output (Fig. 3.1). This particular program is for the solution of a given quadratic equation. These instructions are to be read in order, starting at the top of the deck (i.e., the bottom of the input in Fig. 3.1). It is appropriate not to worry unduly about the details of this illustrative problem; proceed instead to the subsequent outline of the vocabulary and structure of Fortran.

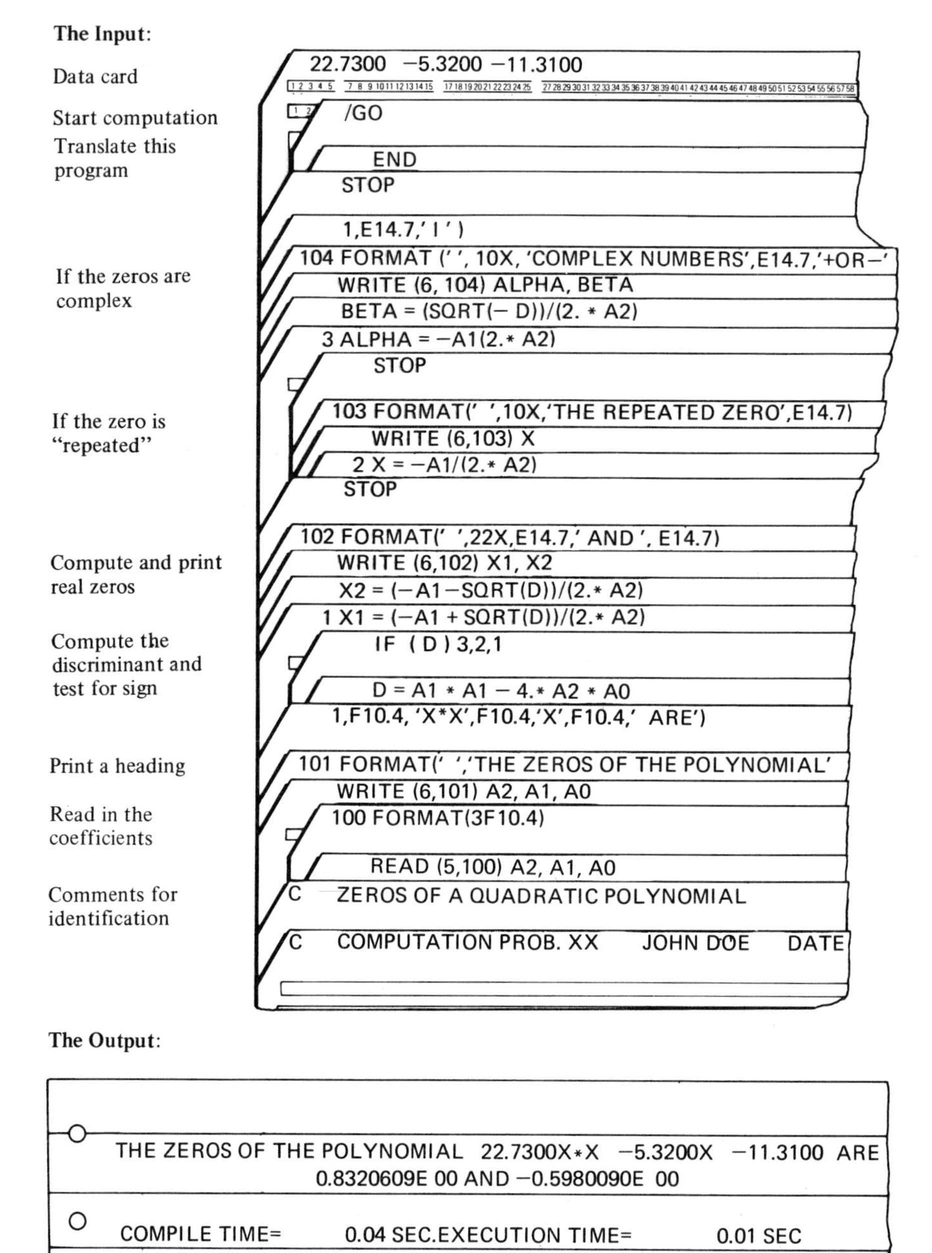
The Input:
Data card
22.7300 −5.3200 −11.3100
Start computation
/GO
Translate this program
END
STOP
1,E14.7,' I ')
If the zeros are complex
104 FORMAT (' ', 10X, 'COMPLEX NUMBERS',E14.7,'+OR−'
WRITE (6, 104) ALPHA, BETA
BETA = (SQRT(− D))/(2. * A2)
3 ALPHA = −A1(2.* A2)
STOP
If the zero is "repeated"
103 FORMAT(' ',10X,'THE REPEATED ZERO',E14.7)
WRITE (6,103) X
2 X = −A1/(2.* A2)
STOP
102 FORMAT(' ',22X,E14.7,' AND ', E14.7)
Compute and print real zeros
WRITE (6,102) X1, X2
X2 = (−A1−SQRT(D))/(2.* A2)
1 X1 = (−A1 + SQRT(D))/(2.* A2)
Compute the discriminant and test for sign
IF (D) 3,2,1
D = A1 * A1 − 4.* A2 * A0
1,F10.4, 'X*X',F10.4,'X',F10.4,' ARE')
Print a heading
101 FORMAT(' ','THE ZEROS OF THE POLYNOMIAL'
WRITE (6,101) A2, A1, A0
Read in the coefficients
100 FORMAT(3F10.4)
READ (5,100) A2, A1, A0
Comments for identification
C ZEROS OF A QUADRATIC POLYNOMIAL
C COMPUTATION PROB. XX JOHN DOE DATE
The Output:
THE ZEROS OF THE POLYNOMIAL 22.7300X*X −5.3200X −11.3100 ARE
0.8320609E 00 AND −0.5980090E 00
COMPILE TIME= 0.04 SEC.EXECUTION TIME= 0.01 SEC

Fig. 3.1

3.2 Number

The Fortran language distinguishes between integers and decimals.

An *integer* can have at most ten digits† and a sign. It must never have a decimal point.

EXAMPLE: +273 −17 27650

If the sign is missing, the integer is assumed to be positive. Integers appear in our computation as given data, or are used for counting or for ordering (their use as subscripts).

A *decimal* may have at most seven significant digits† and a sign. In addition, unless it is zero, its absolute value must be between 10^{-75} and 10^{+75}, and it must have a decimal point.

EXAMPLE: +58.3 −.00007674 27650.

Again, a missing sign indicates a positive number. The notation for the numbers in the above example is called the *F notation*.

In Fortran a decimal can also be expressed in the *E notation*. The numbers above could also be written in the form

```
+5.83E+01  -76.74E-06  27.650E 03
```

or as

```
.583E 02  -.7674E-4  .2765E+05
```

Since the card key-punch machine does not have superscripts, the Fortran notation .583E+02 is an obvious representation of the familiar $(.583)(10^2)$. If the sign after E is omitted, it is assumed that the exponent is positive.

Decimal figures of more than seven significant figures can be read into the machine but are immediately truncated to seven significant figures.

3.3 Variable Names

A *variable name* is a symbol representing a number.

Variable names for integers have at most six alphabetic or numeric characters and must start with *I*, *J*, *K*, *L*, *M*, or *N*.

---

†The number of allowable significant digits differs for the various machines. These limits apply to the IBM 360 machines.

EXAMPLE: N, JOB7, INDEX, L4G,

Variable names for decimals have at most, six alphanumeric characters but must start with a letter other than $I, J, \ldots, N$.

EXAMPLE: FUNCT, XI, VOLT, R2A,

A *subscripted variable name* is a variable name (integer or decimal) followed by an integer or an integer name in parentheses.

EXAMPLE: N(2), VOLT(J), INT(KARD)

The "subscript" in parentheses may also be a linear expression of the form $I * J + K$, *where I and K are specific integers.*

EXAMPLE: X(2*J+3), KOUNT(4*L−5)

Two- and three-dimensional arrays of numbers (double and triple subscripts) may be represented as follows:

EXAMPLE: A(I,J), N(2*J,K)
BAG(I,J,K), JOB(L+2,3*M,N)

Subscripted variable names should be used when a large number of variable-name assignments are made or when it is necessary either to order a set of numbers or to identify the position of a number in an array.

3.4 Statement

A Fortran *statement* is an instruction to the computer. A *nonexecutable statement* supplies descriptive information essential to the compilation of the machine-language version of the Fortran program. An *executable statement* establishes part of the sequence of steps in the calculation.

The physical form of the Fortran statement is a punched card. The following are rules for column usage:

Column	Usage
1	C: for program comments (refer to Section 3.7).
2–5	Blank or a statement number.
6	Blank, or a number 1, 2, . . . , 9 if this card contains a continuation of a statement too long for the preceding one.

7–72	The statement starts on column 7. Blank spaces are ignored.
73–80	Blank or card numbers.

3.5 Dimension Statement

A *DIMENSION statement* is a nonexecutable statement that reserves memory space for subscripted arrays of numbers. Its form is

$$\text{DIMENSION } v_1(n_1), v_2(n_2), \ldots$$

where $v_1, v_2, \ldots$ are variable names and $n_1, n_2, \ldots$ are the number of storage locations to be reserved in machine memory for the corresponding variables.

For two-dimensional arrays, $v(n_3, n_4)$ reserves n_3 times n_4 locations in memory for variable names of the form $v(I, J)$ and their corresponding numbers.

EXAMPLE:

```
DIMENSION A(60),B(10,10),NUM(100)
```

The number in parentheses must be equal to or greater than the number of storage locations needed.

The DIMENSION statement must precede in the program any statements involving these subscripted variables.

3.6 End Statement

An *END statement* is a nonexecutable statement that must appear at the end of the program and before data cards. Its form is simply

END

When this card is read into the machine, the translation of the Fortran program into machine language begins.

3.7 Comments

A card with C punched in column 1 will not be translated by the compiler. If the machine is requested to print the submitted Fortran program, the content of these cards (columns 2–72) will be reproduced. These cards may be used to provide explanations to the reader of the program. They may not be placed immediately before statement subprograms.

EXAMPLE:

```
C   THE FOLLOWING DETERMINES
C   ALPHA AND BETA
```

3.8 Read and Write Statements

In some versions of Fortran, for example Watfiv, there is available *format-free* input/output, a convenience for the beginner. We will outline *formated* input/output statements here, since the *free-field* option is not always available. Formating is not hard to learn, has wide usage, and provides a desirable flexibility.

To submit data to the computer we punch them on *data cards* and, in the program, direct the machine to accept and label them with assigned variable names. The form of this statement is

READ (*i*, *n*) list

"*i*," an integer, is the code for the input device (we will use 5 to represent the punched-card reader); "*n*" is the program statement number of an associated FORMAT statement (the *FORMAT statement* conveys to the computer the form in which the input data are specified on the data card); and "list" is a set of variable names, separated by commas. These names will be assigned to the numbers on the data card in corresponding order.

EXAMPLE:

```
    READ (5,100) A,I,X
100 FORMAT(E14.7,I5,F10.4)
```

For this FORMAT specification, the data card would look as follows:

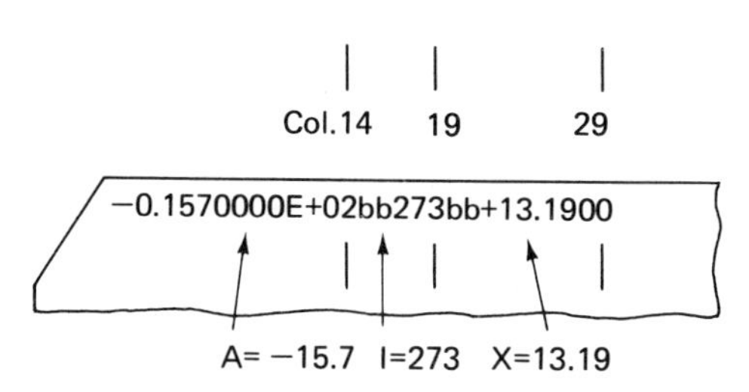

The lowercase letter b is used in these illustrations to emphasize the position of a *blank space* on the card and to clarify spacing.

Data are punched on a card in accordance with the specifications of the format statement. All 80 columns may be used. In Section 3.9 format specifications for numbers are listed. The number of columns used for a given datum is "*w*." These widths are assigned, starting from column 1. Blank spaces on a data card are considered zeros. The data should be punched so that the right-hand column is used in each width.

The output device of the machine is the printer. If, for example, the computer contains numbers with variable names JOB, COST, and WAGE, these will be printed at the command of a WRITE statement, in a specified format:

```
    WRITE (6,101) JOB,COST,WAGE
101 FORMAT('b',15,E14.7,F6.2)
```

For this specification the printed line would look as follows:

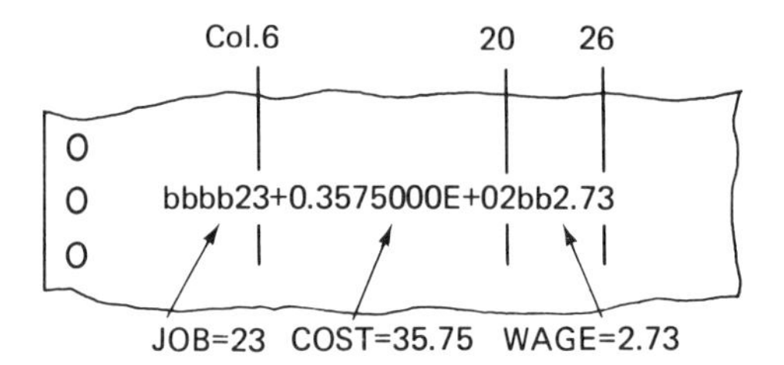

The form of the WRITE statement is

WRITE (*i*, *n*) list

"*i*" is the code for the output device (we will use 6 to represent the printer); "*n*" is the program statement number of an associated FORMAT statement that specifies the form for printing the listed variable values; and "list" is a set of variable names, separated by commas.

3.9 Format Statement

n FORMAT(list)

If this statement is associated with a READ statement, "list" is an ordered set of number descriptions corresponding to the variable names in the READ statement. They may include

Ew.d, Fw.d, Iw

where E and F represent a decimal number

I represents an integer

w is the number of columns allotted to this datum

d is the number of digits to the right of the decimal point

For a FORMAT statement associated with a WRITE statement, "list" contains

1. Number descriptions corresponding to the variable names in the WRITE statement.
2. Carriage controls:

nX	shifts the carriage to skip n columns
///. . . /	k slashes will advance the carriage k lines

The first column will not be printed. If you request that there be printed in column 1:

b	the carriage will single-space
0	the carriage will double-space
1	the carriage will start a new page

3. Literal print outs:

'. . .' The printer reproduces whatever is punched between the apostrophes (the upper symbol on the H key).

It is necessary that "list" be enclosed in parentheses and that all format instructions be separated by commas.

EXAMPLE:

```
      WRITE (6,102)
      WRITE (6,103) FIL,GRID,PLATE
  102 FORMAT('b',18X,'TYPE Z TUBE
     1CHARACTERISTICS'///14X,'FILAMENT',
     29X,'GRID',9X,'PLATE')
  103 FORMAT('b',13x,F10.4,3X,F10.4,3X,F10.4)
```

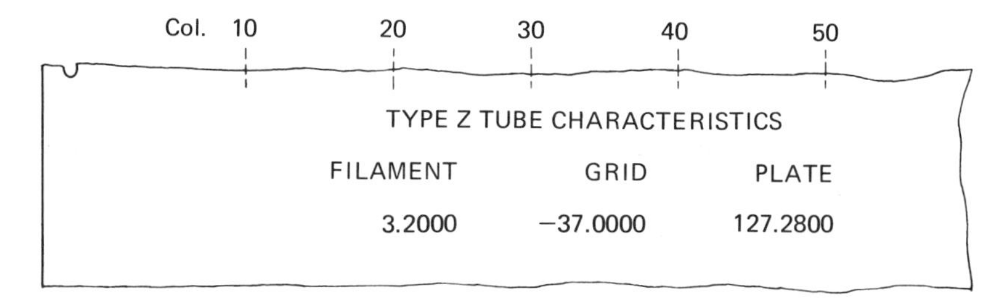

In the above FORMAT statement 103, the same instruction is repeated. A shorter version is permissible:

```
103 FORMAT('b',10X,3(3X,F10.4))
```

In FORMAT specifications for both READ and WRITE statements that involve subscripted variable names, similar shorter versions are possible:

EXAMPLE:

```
    READ (5,104) A(1),A(2),A(3),A(4)
104 FORMAT(F10.6,F10.6,F10.6,F10.6)
```

may be written as

```
    READ (5,105)(A(I),I=1,4)
105 FORMAT(4F10.6)
```

EXAMPLE:

```
    WRITE (6,106)A(1),B(1),A(2),B(2)
106 FORMAT('b',F10.2,E14.7,F10.2,E14.7)
```

may be written as

```
    WRITE (6,107)(A(I),B(I),I=1,2)
107 FORMAT('b',2(F10.4,E14.7))
```

3.10 Expressions

An *expression* is an algebraic combination of constants, variable names, or function subprograms.

The printed symbols for the arithmetic operations are

**	means raised to the power
*, /	means multiply, divide by
+, −	means add, subtract

It is important to observe the order in which operations are to be performed. Parentheses are used in the conventional algebraic sense. In any expression the operations indicated in the inner parentheses are performed first. Within any parentheses the arithmetic operations are performed according to the hierarchy indicated in the above list. Operations at the same level in this

(f)	2,735	(l)	COUNT(3)	(r)	J(A)
(g)	AB*C	(m)	7X	(s)	A(J)
(h)	+.1492E−95	(n)	X7	(t)	Y(2.)
(i)	−9999.E−02	(o)	X(7)	(u)	Y(0)
(j)	$1.98	(p)	JJJ	(v)	Y(−2*1+7)
(k)	KOUNT(−3)	(q)	AJJ	(w)	Y(I*I+7)

3.12.2 Construct statements to read in the data and assign the indicated variable name.

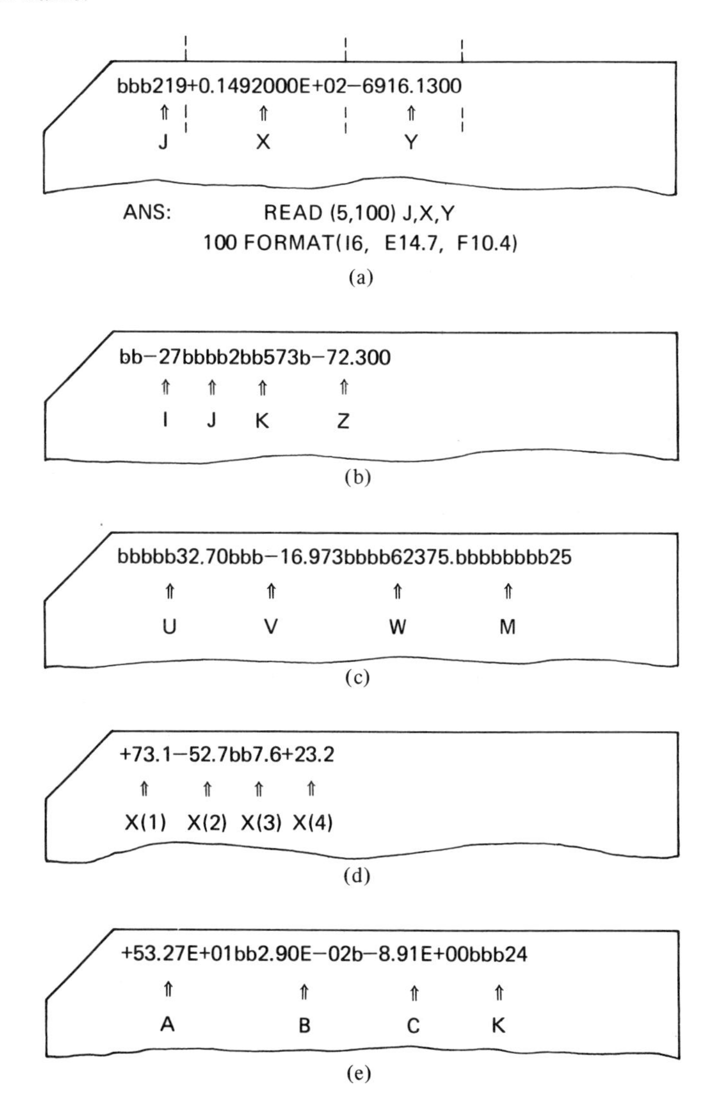

(a)

(b)

(c)

(d)

(e)

3.12.3 Construct statements that will result in the following print outs. The indicated variable names have, in memory, the corresponding numbers.

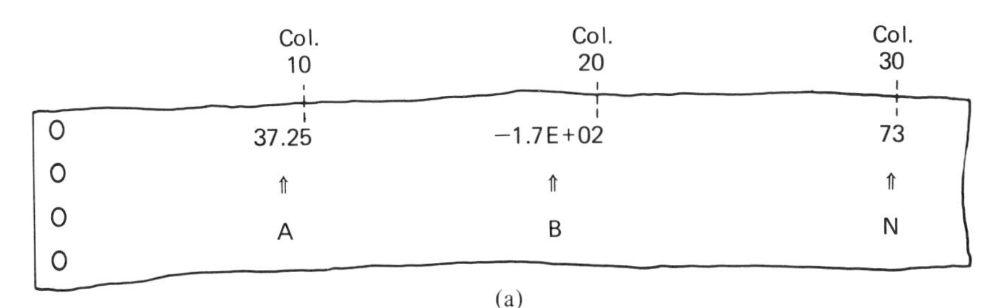

(a)

ANS:

```
      WRITE (6,101) A, B, N
  101 FORMAT ('b', F9.2, E10.1, I10)
```

```
THE FIRST QUADRANT SOLUTION
OF  X*X = Y+1, Y*Y = X+2
IS  X = 1.7106, Y = 1.9262
        ⇑          ⇑
        X          Y
```

(b)

```
           MARY, MARY
        QUITE CONTRARY
  HOW DOES YOUR PROGRAM GROW?
        WITH LOTS OF LOOPS
         IN NESTED GROUPS
  AND NEAT ARRAYS ALL IN A ROW.†
```

(c)

3.12.4 What is the value of each of the following Fortran expressions? For the rules of Fortran arithmetic refer to Section 3.10.

(a)	SQRT(3.**4)/2.	Ans: 4.5
(b)	2.+3./2.**2	Ans: 2.75
(c)	2+3/2**2	Ans: 2

†Reprinted with permission of Datamation®, copyright, Technical Publishing Co. Barrington, Illinois, 60010, 1970.

Example (c) illustrates the rules for division of integers in Fortran.

(d) 2.**3/2.**4
(e) (2.+3./2.)**2
(f) (2+3/2)**2
(g) (−11/3)*2−1
(h) 3./(2.+3.)*2,
(i) 3*5+13/7
(j) (3*5+13)/7
(k) (3.*5.+13.)/7.

3.12.5 Write Fortran statements for

(a) $l \longleftarrow \sqrt{a^2 + b^2} \cos \omega(t - t_1)$

Ans: AL=SQRT(A*A+B*B)*COS(W*(T−T1))

If the assigned variable name were L, the computed value of this expression would be truncated after the decimal point and stored as an integer. If this is to be avoided, it is necessary to assign a variable name for a decimal.

(b) $z \longleftarrow \cfrac{1.}{x + \cfrac{1.}{x + \cfrac{1.}{x}}}$

(c) $r \longleftarrow ab^{x-1}$

(d) $z \longleftarrow z \log_{10} z$

(e) $n_{2j+1} \longleftarrow m_{j+2} l_{j-1}$

(f) $m \longleftarrow \tan^{-1}\left(\frac{a}{a - x}\right)$

(g) $y \longleftarrow \frac{1.}{x - 1.} + \frac{1.}{x - 2.} + \frac{1.}{x - 3.}$

(h) $a_i \longleftarrow \sqrt{b_i^2 + c_{i+1}^2}$

(i) $x \longleftarrow x \tan x$

3.12.6 Each of the following is an incorrect Fortran statement. Write a correct statement based on what you think the intent of the given statement may be.

(a) X=2/3*(X+Y) Ans: X=2./3.*(X+Y)

As stated earlier, in this text we will make expressions consistent in mode.

(b) WAGE=($2.75)HOURS
(c) A+B=C
(d) 2Y=LN(X+1)/(X−1)
(e) P=N+.5**2
(f) INT=PRIN*RATE
(g) X=LOG(X+1)/X−1
(h) Y=EXPON(−X*X/2)
(i) T=SIN(K+2)

3.13 Statement Subprograms

A statement subprogram defines a function to be used in a specific program. This statement has to be placed immediately preceding the first executable statement in the program. It is of the form

$$\text{Name } (a, b, \ldots, z) = \text{expression}$$

Name is a variable name, decimal or integer, identifying the functional value; $a, b, \ldots, z$ are nonsubscripted variable names; and *expression* is an expression involving $a, b, \ldots, z$ and constants.

EXAMPLE:

```
1 RAD(S,A,B,C)=SQRT((S-A)*(S-B)*(S-C)/S)
2 READ (5,200) X,Y,Z
3 S=.5*(X+Y+Z)
  .
  .
  .
4 R=RAD(S,X,Y,Z)
  .
  .
  .
```

Statement 1 is a statement subprogram defining the radius of the circle inscribed in a triangle of sides A, B, and C; statement 2 is the first executable statement of the program; and statement 4 "calls for" the subprogram, evaluates RAD for the computed S and $X = A$, $Y = B$, $Z = C$, and labels the result R. In the "call" statement the variable names do not have to be the same as in the subprogram. Substitution in the statement subprogram is determined only by the order in which the variable names are written.

3.14 Control Statements

3.14.1 Unconditional Transfer

$$\text{GO TO } n$$

This statement means that the next statement in the program to be executed is the one numbered n.

3.14.2 Conditional Transfer

$$\text{IF(expression)}n_1, n_2, n_3$$

This statement means that, at this point in the computation, the value of "expression" is to be computed and tested for sign.

If negative, the next statement to be executed is numbered n_1.
If zero, the next statement to be executed is numbered n_2.
If positive, the next statement to be executed is numbered n_3.

One should be aware that a test for an expected zero value of an expression is sometimes thwarted by a compiler that performs arithmetic such as

$$\begin{aligned} 2. - \sqrt{4.} &= 2.000000 - 1.999999 \\ &= 0.000001 \neq 0 \end{aligned}$$

3.14.3 Stop The execution of the *STOP statement* does the obvious; it stops the calculation. Every computation should have a STOP when the conditions for termination have been met.

3.14.4 The DO Loop

$$\text{DO } n\ i = m_1, m_2, m_3$$

This statement means that the sequence of statements from this one to the one numbered n in the program is to be repeatedly executed, first with integer counter $i = m_1$, again with $i = m_1 + m_3$, again with $i = m_1 + 2m_3$, and so forth until $i > m_2$. Thus for the counter

m_1	is the initial value
m_2	is the upper bound
m_3	is the increment

In this statement i stands for a variable name of an integer.

The last statement in a DO loop must be executable and must not prevent the transfer back to the DO statement. The last statement of the loop cannot be a GO TO, IF, STOP, or another DO statement. If one of these statements is to be used to transfer to a statement outside the loop, and it would normally be listed last, it must be followed, as the last statement of the loop, by

$$n \text{ CONTINUE}$$

Statement number n is the one referred to in the preceding DO statement.

EXAMPLE:

```
     DO 10 I=1,20,1
       READ (5,100) A,B,K
       IF(K-500) 10,10,30
  10 CONTINUE
  30 ...
```

This portion of a program asks that 20 data cards be read. If, however, one of these cards has a value of K greater than 500, transfer is then made to statement 30.

EXAMPLE: Numbers y_3, y_6, y_9 are in memory and we wish to store in memory $2 \sin y_3$, $2 \sin y_6$, $2 \sin y_9$ and print out cumulative sums.

```
     ...
     ...
     SUM=0.
     DO 11 J=1,8,3
       X(J)=2.*SIN(Y(J+2))
       SUM=SUM+X(J)
       WRITE (6,101) SUM
  11 CONTINUE
     ...
```

This program fragment performs the following operations:

j	test	x_j	Sum	Write
1	$1 \leqq 8$	$x_1 = 2 \sin y_3$	x_1	x_1
4	$4 \leqq 8$	$x_4 = 2 \sin y_6$	$x_1 + x_4$	$x_1 + x_4$
7	$7 \leqq 8$	$x_7 = 2 \sin y_9$	$x_1 + x_4 + x_7$	$x_1 + x_4 + x_7$
10	$10 > 8$ →	Transfer to the statement following the last statement of the loop.		

3.15 Program

A *program* is a logically consistent, ordered set of statements. These statements are of the type

1. Read and write
2. Assignment
3. Statement subprograms
4. Control
5. Specification (Dimension, Format)

The executable statements are performed in the order in which they are submitted unless directed otherwise by a control statement.

EXAMPLE: Read in 100 data cards, each containing two numbers u and v, each in F10.4. Print out in E14.7

$$\sqrt{\sum_{i=1}^{100} u_i v_i}.$$

```
      SUM=0.
      DO 1 I=1,100,1
        READ (5,100) U,V
    1 SUM=SUM+U*V
      X=SQRT(SUM)
      WRITE (6,101) X
      STOP
  100 FORMAT(F10.4,F10.4)
  101 FORMAT('b',E14.7)
      END
      [Data cards]
```

3.16 Flow Charts

When the solution of a computational problem requires only a simple pattern of familiar operations, we proceed to write tentative program steps. To attempt immediately the detailed program for a more subtle or involved problem may not be an economical use of our efforts. In such a case it is helpful to look at the broad outline of computational chores: a diagram indicating specific computations and the order in which they are to be performed in programming is called a *flow chart.*

As an example, consider an extension of the illustrative problem at the beginning of this chapter. Instead of solving a single quadratic equation, let a number of data cards be read in. Each card contains the coefficients of a quadratic equation where $A2$, the coefficient of the quadratic term, is not zero. Following the last data card, a card is added having for $A2$ the number zero. This *trailer card* serves as an indicator to the computer that the last data card has been read in. To print out for each equation its coefficients and solution, the following steps could be taken:

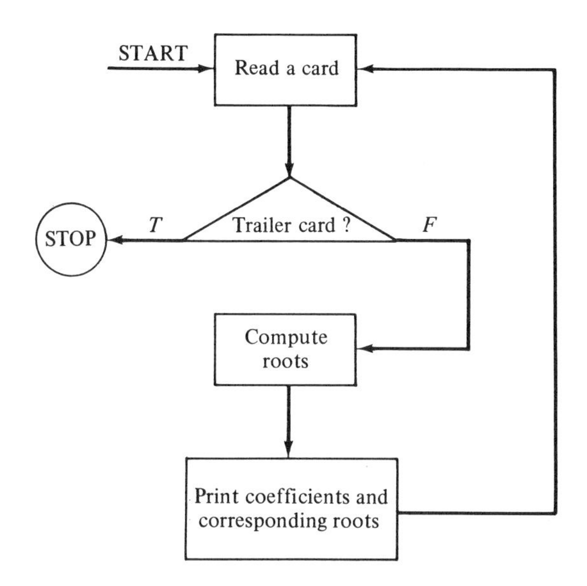

We could be more specific by replacing the box "Compute roots" by

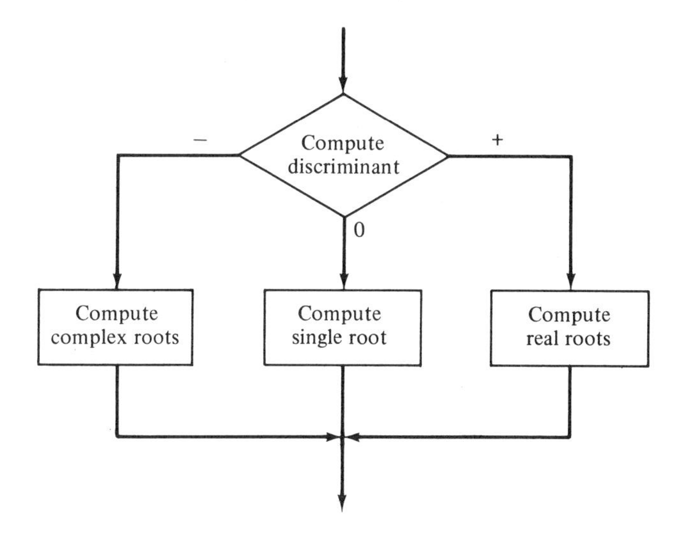

A program for this problem is

```
    6 READ (5,100) A2,A1,A0
      IF(A2)4,5,4
    5 STOP
    4 D=A1*A1-4.*A2*A0
      IF(D)3,2,1
    1 X1=(-A1+SQRT(D))/(2.*A2)
      X2=(-A1-SQRT(D)/(2.*A2)
      WRITE (6,102) A2,A1,A0,X1,X2
      GO TO 6
    2 X=-A1/(2.*A2)
      WRITE (6,103) A2,A1,A0,X
      GO TO 6
    3 ALPHA=-A1/(2.*A2)
      BETA=(SQRT(-D))/(2.*A2)
      WRITE (6,104) A2,A1,A0,ALPHA,BETA
      GO TO 6
  100)
  102 \
      }Appropriate format statements
  103 /
  104)
      END
```

Flow charts are not unique to computational or data processing problems. When an organization grows to contain a large number of interrelated subgroups, an organizational chart is a helpful device to simply explain these relationships. If the product of a corporation, for instance a steam turbine, requires a large number of interrelated and time dependent actions, a production flow chart is a helpful graphic device to display this sequence of operations. Neither the organizational chart nor the production flow chart is necessary at the corner shoe repair shop even though they could be made.

When a computation involves a large number of interrelated sequenced operations, a flow chart is an asset in planning computations and searching for improved strategies. This should rarely be the case for the problems of this book. However, the following flow chart may suggest a relevant program for an evening when there's nothing really good on TV.

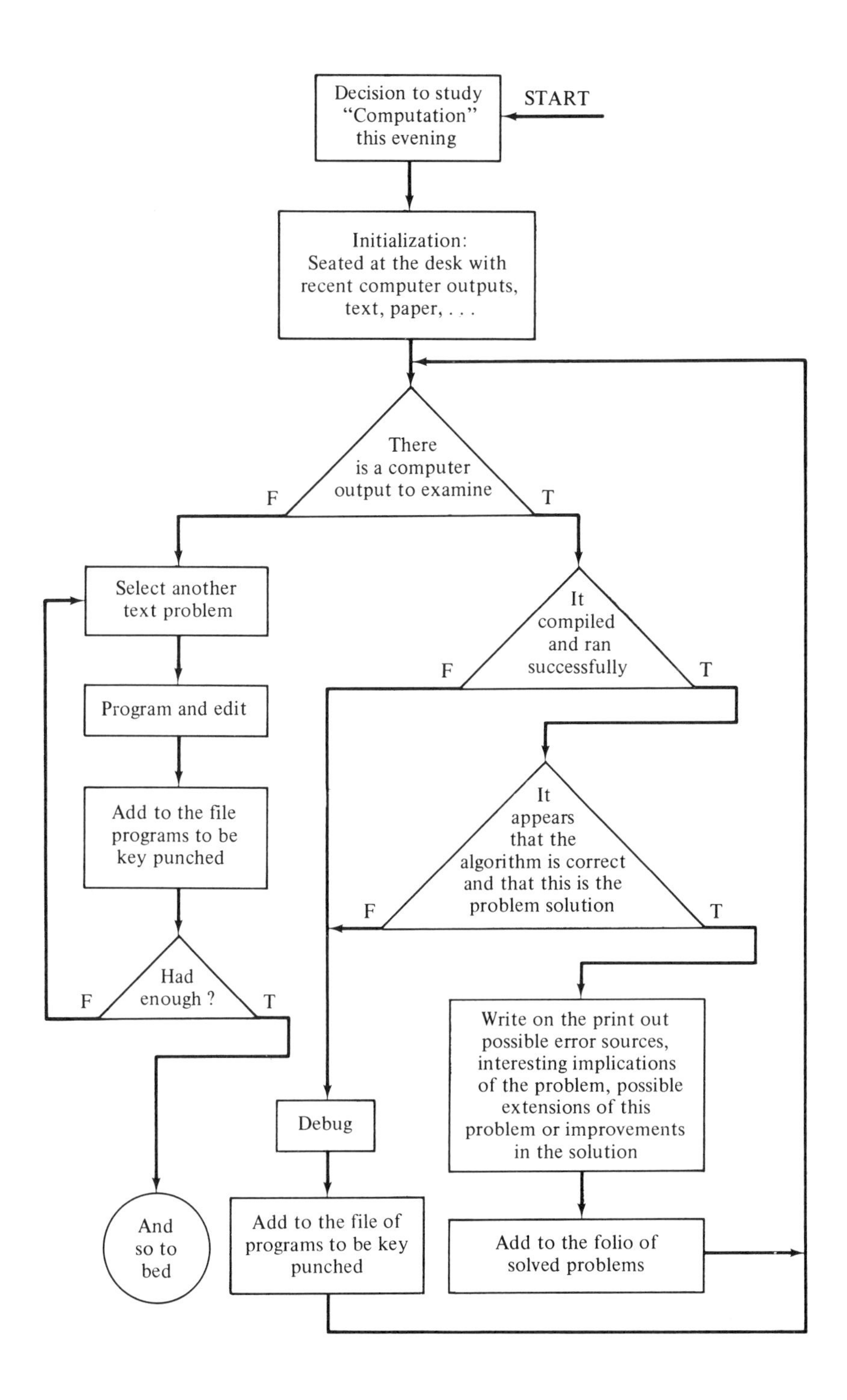
START
Decision to study "Computation" this evening
Initialization: Seated at the desk with recent computer outputs, text, paper, . . .
There is a computer output to examine
F
T
Select another text problem
Program and edit
Add to the file programs to be key punched
Had enough ?
F
T
And so to bed
It compiled and ran successfully
F
T
It appears that the algorithm is correct and that this is the problem solution
F
T
Debug
Add to the file of programs to be key punched
Write on the print out possible error sources, interesting implications of the problem, possible extensions of this problem or improvements in the solution
Add to the folio of solved problems

3.17 Examples in Programming

3.17.1 Read in 50 numbers, punched one to a card, each in E14.7. Assign them the variable names $A(1), A(2), \ldots, A(50)$.

```
    DIMENSION A(50)          An allotment of storage space
    DO 10 K=1,50,1           } A DO loop involving 50 READ
 10 READ (5,100) A(K)        } instructions
100 FORMAT(E14.7)
    ||
    ||
    ||
```

3.17.2 Read in one card specifying N in I3, then read in that many cards, each containing two numbers, each in F10.4. Label these $X(1), X(2), X(3), \ldots$.

```
    DIMENSION X(1000)             Allotment of storage space
    READ (5,100) N                made with some knowledge
                                  of an upper bound for 2N
    M=2*N
                                  } To be assured that this assigns
    DO 10 K=1,M,2                 } the required variable names,
 10 READ (5,101) X(K),X(K+1)      } mentally run through several
                                  } cycles of the loop.
100 FORMAT(13)
101 FORMAT(F10.4,F10.4)
    ||
    ||
    ||
```

3.17.3 Read in data cards, each containing one number in E14.7, until a trailer card indicating the end of the data is read. Print out the number of data and their sum.

In this problem there is no need to save data and thus no need for subscripted variable names. If upper or lower bounds of the data are known, a trailer card contains a number identifiable as not a member of the data set.

```
    J=0                       } Initialization of the counter,
    SUM=0.                    } and the sum
  2 READ (5,100) X            { Test for the trailer card.
    IF (X-1.0E+10) 1,3,3      { Here it is assumed that 10^10
  1 SUM+SUM+X                 { is greater than any datum. The
    J=J+1                     { trailer card could contain, for
    GO TO 2                   { instance, 5.E+10.
  3 WRITE (6,101) J,SUM
    STOP
```

```
100 FORMAT(E14.7)
101 FORMAT('b',I10,3X,E14.7)
    END
```

3.17.4 Read in data that are punched eight numbers per card, each number in F10.4. Label them

$$X(1),\ Y(1),\ X(2),\ Y(2),\ X(3),\ Y(3),\ X(4),\ Y(4),\ X(5),\ Y(5),\ X(6),\ \ldots$$

These number pairs may represent coordinates of points in a plane. Assume that $+9999$ is greater than any abscissa in the data set. Let this number be a trailer indicating the last data. For instance, if 58 coordinate number pairs are to be read in, the trailer X will be the third X or the fifth number on the fifteenth card.

Count the number of points read in, and after they are labeled and stored, compute $\sum_{i=1}^{n} x_i y_i$.

```
    DIMENSION X(200),Y(200)             Storage allotment based on
    J=1                                 prior knowledge of an
  2 K=J+3                               upper bound on the number
    READ (5,100) ((X(I),Y(I),I=J,K)     of points
    DO 10 L=J,K,1                       } Test for trailer X
      IF (X(L)-9999.) 10,3,3            } for each card read
 10 CONTINUE                            } in.
    J=J+4
    GO TO 2
  3 N=L-1                               Number of points read in
    SUM=0.
    DO 20 I=1,N,1
 20 SUM=SUM+X(I)*Y(I)                   The final value of SUM is
    WRITE (6,101) N,SUM                 Σ(i=1..n) x_i y_i.
    STOP
100 FORMAT(8F10.4)
101 FORMAT('b',I10,3X,E14.7)
    END
```

3.17.5 Generate the array $A(I, J) = I + J$, where $I = 1, 2, \ldots, 10$ and $J = 1, 2, \ldots, 10$. Print out the triangular array $A(I, J)$, where $J \leq I$:

$$\begin{array}{l} A(1,1) \\ A(2,1),\ A(2,2) \\ A(3,1),\ A(3,2),\ A(3,3) \\ \vdots \\ A(10,1),\ \ldots,\qquad A(10,10) \end{array}$$

```
      DIMENSION A(10,10)
      DO 10 I=1,10,1                    The square matrix is generated
        DO 20 J=1,10,1                  row by row.
 20     A(I,J)=I+J
 10   CONTINUE
      DO 30 I=1,10,1                    Writing out the lower triangular
        K=1                             element
 30   WRITE (6,100) (A(I,J),J=1,K)
      STOP
100   FORMAT('b',10(3x,F5.0))           Note that unused format is
      END                               skipped.
```

3.17.6 Given that $X(I)$ and $Y(I)$, where $I = 1, 2, \ldots, 100$, are in the machine memory. Find the sum of the absolute value of the X's and the product of the Y's.

```
      ||
      ||
      ||
      SUM=0.                            Initialization of the sum and
      PROD=1.                           product
      DO 10 I=1,100,1                   Loop generating a sequence of
        SUM=SUM+ABS(X(I))               partial sums and products
 10   PROD=PROD*Y(I)
      WRITE (6,100) SUM,PROD            Printing out the last elements of
      ||                                these two sequences
      ||
      ||
```

3.17.7 Given that $X(I)$, where $I = 1, 2, \ldots, 100$ represents numbers in memory. Find the largest of these numbers.

```
      ||
      ||
      ||
      A=X(1)                            Initial candidate for the largest
                                        number
      DO 10 I=2,100,1
        IF (X(I)-A)10,1,1               In this sequence of tests A
      1 A=X(I)                          represents the largest of those
 10   CONTINUE                          tested.
      WRITE (6,100) A
      ||
      ||
      ||
```

3.17.8 Given that Z and $X(I)$, where $I = 1, 2, \ldots, 100$ represent numbers in memory. Find a number $X(I)$ differing from Z, in absolute value, by the least amount.

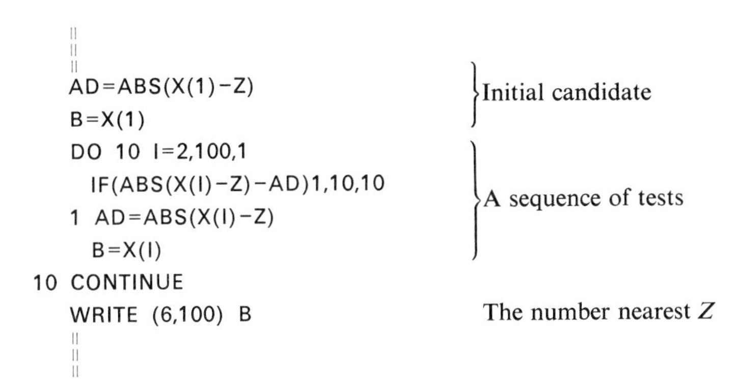

```
     ||
     ||
     ||
     AD=ABS(X(1)-Z)              } Initial candidate
     B=X(1)                      }
     DO 10 I=2,100,1             }
       IF(ABS(X(I)-Z)-AD)1,10,10 }
     1 AD=ABS(X(I)-Z)            } A sequence of tests
       B=X(I)                    }
  10 CONTINUE
     WRITE (6,100) B               The number nearest Z
     ||
     ||
     ||
```

3.17.9 Given A, B, and C, representing numbers in memory.

(a) If $A - B \leqq 0$ *and* $|C| - B < 0$, transfer to statement 10, otherwise to statement 20.

```
  ||
  ||
  ||
  IF(A-B)1,1,20
1 IF(ABS(C)-B)10,20,20
  ||
  ||
  ||
```

(b) If $A + B > 0$ *or* $C^2 - B \geqq 5$, transfer to statement 30, otherwise to statement 40.

```
  ||
  ||
  ||
  IF(A+B)1,1,30
1 IF(C*C-B)40,30,30
  ||
  ||
  ||
```

(c) If $A > 0$ *or both* $B < 0$ *and* $C > 5$ transfer to statement 50, otherwise to statement 60.

```
  ||
  ||
  ||
  IF(A)1,1,50
1 IF(B)2,60,60
2 IF(C-5)60,60,50
  ||
  ||
  ||
```

3.17.10 Tabulate $f(x) = x^2$ for $x = [0, .1, 1.0]$ (or $x = .0, .1, .2, \ldots, 1.0$ or $x = n(.1)$, where $n = 0, 1, 2, \ldots, 10$).

Of the following two programs for this tabulation, the one on the left contains a statement subroutine.

```
     F(X)=X*X
     X=0.
     DO 10 I=1,11,1
       G=F(X)
       WRITE (6,100) X,G
       X=X+.1
  10 CONTINUE
     STOP
 100 FORMAT('b'F4.1,3X,F10.7)
     END
```

```
     X=-.1
   1 X=X+.1
     F=X*X
     WRITE (6,100) X,F
     IF(X-1.0)1,1,2
   2 STOP
 100 FORMAT('b'F4.1,3X,F10.7)
     END
```

3.17.11 Is 733 a prime number? (This example illustrates a use for the Fortran division of integers.)

```
     K=733
     A=K
     N=SQRT(A)+1.
     DO 10 I=2,N,1
       IF(K-K/I*I)10,1,10
  10 CONTINUE
     WRITE (6,100)
     STOP
   1 WRITE (6,101)
     STOP
 100 FORMAT('b733 IS A PRIME NUMBER')
 101 FORMAT('b733 IS NOT A PRIME NUMBER')
     END
```

A=733.
N=28
For the first cycle
733/2=366
733/2*2=732
733−733/2*2=1

3.17.12 Write a program corresponding to the following flow chart:

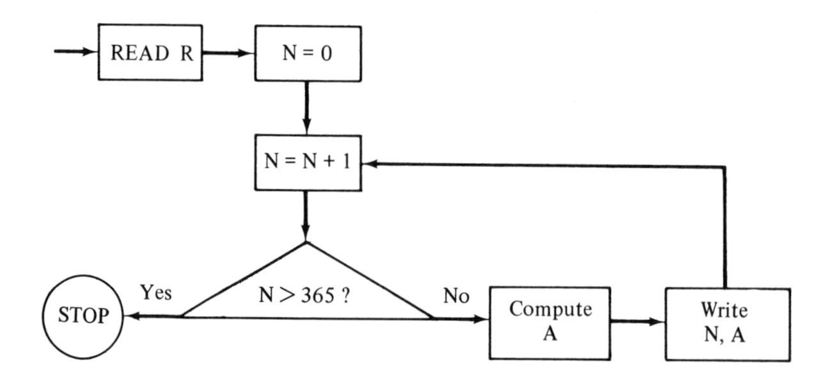

This program is to tabulate the interest to be paid on \$1.00 invested for N days at an annual rate of R, interest compounded daily.

$$A = \left(1. + \frac{R}{365.}\right)^N - 1.$$

```
     READ (5,100) R
     N=0
   3 N=N+1
     IF(N-365)2,2,1
   1 STOP
   2 A=(1.+R/365.)**N-1.
     WRITE (6,101) N,A
     GO TO 3
 100}
 101}FORMAT STATEMENTS
     END
```

```
     READ (5,100) R
     B=1.+R/365.
     AA=1.
     N=0
   3 N=N+1
     IF(N-365)2,2,1
   1 STOP
   2 AA=AA*B
     A=AA-1.
     WRITE (6,101) N,A
     GO TO 3
 100}
 101}FORMAT STATEMENTS
     END
```

For each of these two programs count the number of multiplications in the repeated execution of statement 2.

3.18 Problems

3.18.1 Given 20 data cards, each containing three numbers in F10.4. Read these numbers in and label them $X(1), X(2), \ldots, X(60)$.

3.18.2 Read in the coordinates of 50 three-space points that are punched two points (six numbers) per card, each number in F10.4. Label them $X(1)$, $Y(1)$, $Z(1)$, $X(2)$, $Y(2)$, $\ldots$, $Z(50)$.

3.18.3 Given $A(I)$ and $B(I)$, where $I = 1, 2, \ldots, 20$, in machine memory. Write statements to print these numbers in tabular form with the headings

```
O                         TABULATION OF DATA

O               I              A                  B


O               XX        ± XXX.XXXX        ± XXX.XXXX

O               XX        ± XXX.XXXX        ± XXX.XXXX
```

3.18.4 On one card are four numbers, each in F10.4. They represent (x_1, y_1) and (x_2, y_2), coordinates of two points on a plane. Read in these numbers and print out: "The line containing (__, __) and (__, __) is $y =$ __$x +$ __," with appropriate numbers appearing in place of the dashes. Make provision for the proper response if the two given points have the same abscissa.

3.18.5 L and M represent positive integers in machine memory. Print out the logarithm of the larger. If they are equal, print out their logarithm.

3.18.6 X and Y each represent a number in machine memory. Write program segments that will transfer to statement 50 if

(a) Both $X \leqq 5$ and $Y < 10$.
(b) Either $X \leqq 5$ or $Y < 10$.

3.18.7 X, Y, and Z each represent a number in machine memory. Write program segments that will transfer to statement 60 if

(a) $X < 0$ and either $Y > 0$ or $Z \geqq 0$.
(b) $X < 0$ or both $Y > 0$ and $Z \geqq 0$.

3.18.8 A, B, and C each represent a number in machine memory. Print out A if

$$A < B \leqq C = 100$$

Otherwise, print out B.

3.18.9 Read in the coordinates of three points in the plane (x_1, y_1), (x_2, y_2), (x_3, y_3), punched on one card, each number in F10.4. Print out the equation of the line containing the first two points in the form

$$Ax + By + C = 0$$

and also the distance from the third point to this line:

$$D = \frac{|Ax_3 + By_3 + C|}{\sqrt{A^2 + B^2}}$$

Make sure that your program will cope with the case for which $x_1 = x_2$.

3.18.10 For the circle of radius 2, with the center at the origin, print out the coordinates of all points in the first quadrant for which $x = n(.1)$, where $n = 1, 2, 3, \ldots,$

3.18.11 Read in the coordinates of 52 points, 4 points per data card (8 numbers) each in F10.4. Print out the centroid of these points $(\bar{x}, \bar{y})$, where

$$\bar{x} = \frac{1}{n}\sum_{i=1}^{n} x_i, \quad \bar{y} = \frac{1}{n}\sum_{i=1}^{n} y_i$$

3.18.12 Read in a deck of data cards, each containing a positive number in E14.7. Let a trailer card contain a negative number. Print out n, the number of data, and the arithmetic and geometric means:

$$A = \frac{1}{n}\sum_{i=1}^{n} x_i, \quad G = \sqrt[n]{\prod_{i=1}^{n} x_i}$$

3.18.13 Print out the first 20 of the sequence of products

$$y_n = \frac{1}{1-\frac{1}{2}} \cdot \frac{1}{1-\frac{1}{2^2}} \cdot \frac{1}{1-\frac{1}{2^3}} \cdots \frac{1}{1-\frac{1}{2^n}}, \qquad n = 1, 2, 3, \ldots$$

3.18.14 An array of numbers $C(I, J)$, where $I = 1, 2, \ldots, 6$ and $J = 1, 2, \ldots, 5$, are punched on six cards, five to a card, each in F10.4. Read in these numbers and print out the sum of the interior numbers (those not on the first or last row or column).

3.18.15 A number E and an 8-by-8 array $A(I, J)$ are in machine memory. Write a program fragment that goes to STOP if the sum of the absolute values of the elements below the diagonal elements is less than E.

3.18.16 Given that numbers labeled X and $A(J)$, where $J = 1, 2, \ldots, 100$, are in machine memory. Print out the number of elements in this array that are greater than X.

3.18.17 Numbers labeled $B(I)$, where $I = 1, 2, \ldots, 100$ are in machine memory. Print out the number of elements in this array in each of the following intervals:

$$[0., 200.), [200., 800.], (800., 1000.]$$

3.18.18 Read in a set of positive numbers punched one to a card, each in E14.7. Assume that each number is less than 1.E+20. Print out their arithmetic mean and their range (the greatest minus the least). In this program no subscripts are needed.

3.18.19 Read in a set of numbers (less than 100) punched one to a card, each in E14.7. Assume that each number is less than 1.E+20. Print out the number closest to their arithmetic mean. In this program it will be necessary to use subscripted variable names so that, after the mean is computed, each can then be compared to the mean.

3.18.20 Read in elements of a ten-row, five-column array a_{ij}. These numbers are punched by rows, five to a card. Print out

$$\text{(a)} \quad \sum_{j=1}^{5} |a_{ij}|, \quad i = 1, 2, 3, \ldots, 10$$

$$\text{(b)} \quad \sum_{i=1}^{10} |a_{ij}|, \quad j = 1, 2, 3, 4, 5$$

In each of these problems is it necessery to first read in the entire array with subscripted labels and then do the computation and print outs?

3.18.21 Read in (x, y), the rectangular coordinates of a point, and print out *the* corresponding polar coordinates (r, θ), where $-(\pi/2) < \theta \leqq (\pi/2)$, $-\infty < r < \infty$.

3.18.22 Write a program that will print out the correct statement:

"403 is a prime number"

or

"403 is not a prime number"

3.18.23 List all prime numbers less than 100.

3.18.24 List all prime numbers less than 100 that differ by 2.

3.18.25 Read in two integers and print out the correct statement:

"— — — is divisible by — — —"

or

"— — — is not divisible by — — —"

In each statement the first and second integers read in are printed in place of the dashes.

3.18.26 Read in a number in F10.4 and print it out in F9.3:

(a) Truncated to three decimal places.

(b) Rounded to three decimal places.

For part (b),

for	2.5633	2.5637
add	.0005	.0005
	2.5638	2.5642
multiply by 10^3	2 563.8	2 564.2
integer function	2 563	2 564
multiply by 10^{-3}	2.563	2.564

4

FORMULA EVALUATION

To exercise our programming abilities this chapter contains a variety of computational situations. The first 14 ask us to evaluate formulas, construct tables, and tabulate functions. The remaining problems are concerned with convergent sequences. In addition to extending programming skills, the solution of these problems should promote added insight into sundry aspects of elementary analysis.

4.1 An Interest Table

4.1.1 The value of an account into which \$1.00 is deposited at the beginning of years $n = 1, 2, 3, \ldots$ and to which interest is added annually at rate r is

$$V = \frac{(1 + r)^n - 1}{r}$$

Print out a tabulation of V in the form

```
O                         VALUE OF AN ACCOUNT

O           ONE DOLLAR DEPOSITED AT THE BEGINNING OF
            EACH YEAR N, ANNUAL INTEREST AT RATE R

O    N\R          0.03          0.04          0.05          0.06
      1          XX.XXXX       XX.XXXX        . . .
O
      2          XX.XXXX
                   ||
O     3            ||
      |            ||
O     |
      |
      |
O    30
```

4.1.2 A house is to be purchased for \$30,000 with a down payment of X dollars. Each of 20 annual payments includes \$750 for fixed charges (taxes, sewer rent, etc.), \$$(30{,}000 - X)/20$ for amortization, and interest at rate R on the unpaid balance.

Tabulate for $X = 0, 5{,}000, 10{,}000, 15{,}000$ the payment due at the end of year $n = 1, 2, 3, \ldots, 20$ for interest rates $R = .06, .07, .08, .09$.

4.2 Solution of a Triangle

One of the problems of triangle solving is: Given the three sides a, b, c, find the corresponding angles α, β, γ.

Relevant relationships between sides and angles illustrated in Figure 4.1 are

$$s = \tfrac{1}{2}(a + b + c), \qquad \text{half the perimeter}$$

$$r = \sqrt{(s-a)(s-b)(s-c)/s}, \qquad \text{radius of the inscribed circle}$$

$$\alpha = 2\tan^{-1}\left(\frac{r}{s-a}\right), \quad \beta = 2\tan^{-1}\left(\frac{r}{s-b}\right), \quad \gamma = 2\tan^{-1}\left(\frac{r}{s-c}\right)$$

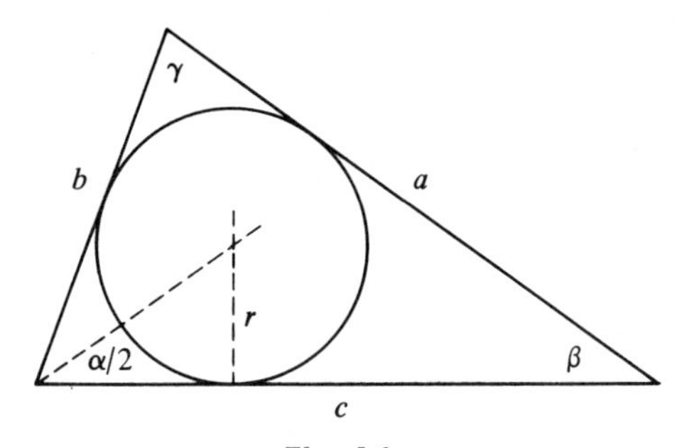

Fig. 4.1

It is possible to read in values of a, b, and c for which α, β, and γ are not computable. Is the existence condition "any side is less than the sum of the other two sides" equivalent to $(s - a)(s - b)(s - c)/s > 0$?

4.2.1 Select a set of values for a, b, and c. Write a program to print out the angles α, β, and γ in degrees. Also print out $E = |\alpha + \beta + \gamma - 180|$.

The output should be of the form

```
O
        FOR THE TRIANGLE WHOSE SIDES ARE
O
        A =          B =           C =
O

O
        THE CORRESPONDING ANGLES ARE
O
        ALPHA =      BETA =        GAMMA =          ERROR =
```

or

```
O
        NO TRIANGLE IS FORMED IF
O
        A =          B =           C =
O
```

4.2.2 Alter the preceding program to write out under a single heading the solutions of several triangles. Punch the lengths of the three sides of each triangle on a card and identify a trailer card with a side given as negative.

4.3 An Empirical Formula

Manning's formula for the velocity of water flowing in a ditch is

$$V = \frac{1.49}{c} R^{2/3} S^{1/2}$$

where V = velocity, in feet per second
S = slope of the ditch

$$R = \frac{A}{P}, \quad A = \text{area of the cross section of water, in square feet}$$

$$P = \text{length of the wetted perimeter of the ditch, in feet}$$

$$c = \text{a coefficient that depends on the roughness of the ditch surface}$$

(Examples of possible variations in c are as follows: concrete, $c = .012$; earth, $c = .020$; grass, $c = .023$; and rubble, $c = .025$.)

4.3.1 For a ditch of trapezoidal section, with base b and "side angle" θ as indicated in Figure 4.2, read in $b = 4$ ft, $\theta = 30°$, $S = .02$, and $c = .020$. Print out data to plot curves showing velocity, and also volume flowing past a section as a function of depth h, $0 \leqq h \leqq 2$. Draw these curves.

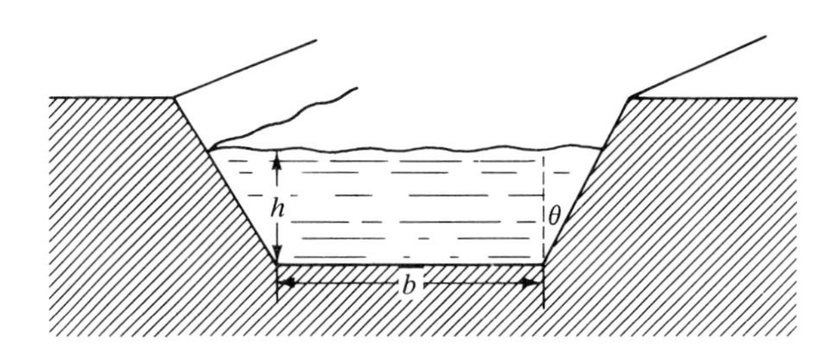

Fig. 4.2

4.4 Mean and Standard Deviation

Given the set of numbers $x_1, x_2, x_3, \ldots, x_n$, the *mean* of this set is

$$\bar{x} = \frac{1}{n} \sum_{i=1}^{n} x_i$$

The *variance* of the set is

$$S^2 = \frac{1}{n-1} \sum_{i=1}^{n} (x_i - \bar{x})^2 = \frac{1}{n-1} \left(\sum_{i=1}^{n} x^2 - n\bar{x}^2 \right)$$

The second formula provides programming conveniences but may result in a loss of accuracy because of the possible small difference of two large numbers. The *standard deviation* is $S = \sqrt{S^2}$.

4.4.1 Let each student in the class independently estimate the age of some person known to all students. Find the mean and standard deviation of this set of numbers. Use a trailer card to indicate the last datum. (Refer to problem 3.17.3.)

4.4.2 Find the square root of the mean value of $\sin^2 x$ over $0 \leq x \leq \pi/2$. Use values of $\sin^2 x$ for $x = n\pi/50$, $n = 1, 2, 3, \ldots, 25$. Also do this analytically.

4.4.3 Toss 5 coins 20 times. Record the number of heads on each toss. Compute the mean and variance of these 20 numbers. Pool your results with those of your classmates. For the 20 c trials (c being the number in the pool) compute $\bar{x}$ and s^2.

As the number of trials increases, the resulting $\bar{x}$ and s^2 should approach

$$\bar{x} \longrightarrow \tfrac{5}{2}, \quad s^2 \longrightarrow \tfrac{5}{4}$$

4.5 Even and Odd Functions

An even function has the property

$$\mathrm{E}(x) = \mathrm{E}(-x)$$

An odd function has the property

$$\mathrm{Od}(x) = -\mathrm{Od}(-x)$$

The graph of an even function is symmetric with respect to the y-axis and the graph of an odd function is symmetric with respect to the origin. Some examples are

Even functions	*Odd functions*
x^n, n an even integer	x^m, m an odd integer
$\cos x$	$\sin x$
$x \sin x$	$\tan x$
$1/(1 - x^2)$	$x^m \cos x$, m an odd integer

An examination of this list suggests some generalizations concerning sums and products of even and odd functions; for example: Is the product (or quotient) of an even and an odd function even, or odd, or neither?

Any function defined over the symmetric interval $-a \leqq x \leqq a$ can be uniquely expressed as the sum of an even function and an odd function,

$$f(x) = \frac{f(x) + f(-x)}{2} + \frac{f(x) - f(-x)}{2}$$

On the right-hand side of the above identity the first term is even and the second odd. Following are some examples:

$$\frac{1}{1-x} = \frac{1}{1-x^2} + \frac{x}{1-x^2}$$

$$\sin(ax+b) = \sin b \cos ax + \cos b \sin ax$$

$$e^x = \frac{e^x + e^{-x}}{2} + \frac{e^x - e^{-x}}{2} = \cosh x + \sinh x$$

4.5.1 Use a statement subroutine to define $\sqrt{x+1}$, and in terms of this tabulate the even and odd components for $x = -1. + n(.1)$, where $n = 0, 1, 2, \ldots, 20$. Draw the graph of $\sqrt{x+1}$ and its even and odd components over $[-1, 1]$.

4.5.2 Write a program that will read in A, N, and functional values $Y(X)$ corresponding to $X = -A + K(A/N)$, where $K = 0, 1, 2, \ldots, 2N$. Print out the even and odd components of $Y(X)$. If these components are properly labeled, is it necessary to print out the component values for $-A \leqq x < 0$?

4.5.3 Find formulas for the even and odd components of

$$f(x) = \frac{x+1}{x^2+x+1}$$

and

$$g(x) = \frac{x^2+x+1}{x+1}$$

Consider the tabulation of the even and odd components of each of these functions over an interval that includes $x = 1$. Can one program be used for both tabulations by merely changing a statement subroutine card?

4.6 Trigonometric Functions

A study of trigonometry begins with a definition of six functions, not all independent. A definition of these functions most useful to the beginning student is in terms of the coordinates of a point on the unit circle as shown in Figure 4.3. Let x measure the length of the arc on the circle $u^2 + v^2 = 1$

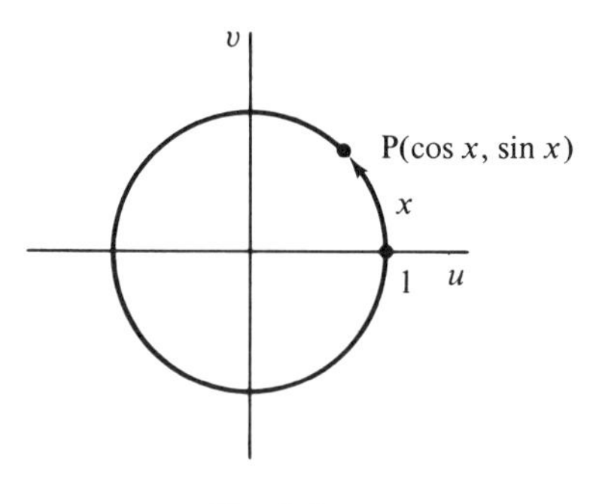

Fig. 4.3

from (1, 0) counterclockwise to P. The functions $\cos x$ and $\sin x$ are defined as the coordinates of P. Four other functions are defined algebraically in terms of $\sin x$ and $\cos x$:

$$\tan x = \frac{\sin x}{\cos x}, \quad \cot x = \frac{1}{\tan x}, \quad \sec x = \frac{1}{\cos x}, \quad \csc x = \frac{1}{\sin x}$$

There are numerous interesting and useful relationships between these functions. For example;

$$\cos^2 x + \sin^2 x = 1, \qquad \text{a } \textit{Pythagorean identity}$$

$$\cos(x + x_1) = \cos x \cos x_1 - \sin x \sin x_1, \qquad \text{an } \textit{addition identiy}$$

$$T\left(x + n\frac{\pi}{2}\right) = \pm \begin{cases} T(x) \\ T_c(x), & \text{the } \textit{cofunction} \text{ of } T(x) \end{cases}$$

These are the *reduction identities* of trigonometry. The selection of the appropriate right-hand side depends on n and T, the trigonometric function. (A high school course in trigonometry presents proof of these relationships and many others that can be derived from them.)

Two useful trigonometric identities are

$$a \cos \omega t + b \sin \omega t = \sqrt{a^2 + b^2} \cos \omega(t - t_1), \quad t_1 = \frac{1}{\omega} \tan^{-1} \frac{b}{a} \tag{1}$$

$$= \sqrt{a^2 + b^2} \sin \omega(t + t_2), \quad t_2 = \frac{1}{\omega} \tan^{-1} \frac{a}{b} \tag{2}$$

These are essentially the basic addition identities. Considering Figure 4.4,

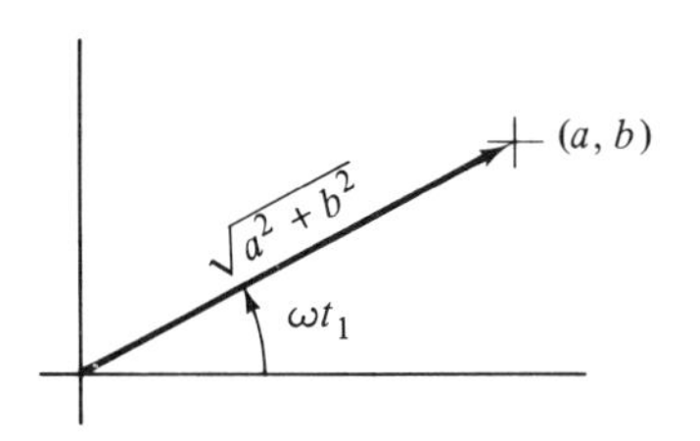

Fig. 4.4

$$a \cos \omega t + b \sin \omega t = \sqrt{a^2 + b^2}\left(\frac{a}{\sqrt{a^2 + b^2}} \cos \omega t + \frac{b}{\sqrt{a^2 + b^2}} \sin \omega t\right)$$

$$= \sqrt{a^2 + b^2}\,(\cos \omega t \cos \omega t_1 + \sin \omega t \sin \omega t_1)$$

$$= \sqrt{a^2 + b^2} \cos \omega(t - t_1)$$

This cosine function is said to have *amplitude* $\sqrt{a^2 + b^2}$, period $2\pi/\omega$, and displacement t_1.

4.6.1 Derive the second identity expressing $a \cos \omega t + b \sin \omega t$ as the sine of a linear function of t. Show that these identities are independent of the signs of a and b. What is the relationship between ωt_1 and ωt_2?

4.6.2 Print, with appropriate problem identification and column headings, tabulations of the following:

(a) $24 \cos t/8 + 7 \sin t/8$, where $t = 0, 2, 4, 6, \ldots, 50$.
(b) The same function in the form $\sqrt{a^2 + b^2} \cos \omega(t - t_1)$.
(c) The difference between corresponding values of (a) and (b).

If entries in column (c) are other than zero, what may be the source of these differences? Draw the graph of the function and state its amplitude, period, and displacement in t with respect to $\sqrt{a^2 + b^2} \cos \omega t$.

4.6.3 In the preceding problem, the sum of a sine and cosine function, each having the same period, was expressed as either a sine or cosine function. As an example of a sum of sines and cosines not having the same period, consider

$$f(x) = \sin x - \cos 2x$$

Tabulate x, $\sin x$, $-\cos 2x$, and $f(x)$ for $x = n \cdot \pi/20$, where $n = 0, 1, 2, \ldots, 40$. Draw the graphs of $\sin x$, $-\cos 2x$, and $f(x)$.

Add to this program so that in addition to the tabulation of these functions, you will print out, labeled as AMPLITUDE, the absolute value of the largest tabulated deviation of $f(x)$ from zero, and the corresponding value of x.

4.6.4 The relationship

$$\sin(\cos x) < \cos(\sin x)$$

may not be easy to prove (see E. W. Hobson, *Plane Trigonometry*, New York, Cambridge University Press, 1939, p. 136). However, we can demonstrate this inequality beyond all reasonable doubt by tabulating both these functions over one complete period.

First evaluate $\sin(\cos x)$ and $\cos(\sin x)$ for $x = n(.2)$, where $n = 0, 1, 2, \ldots, 16$. Then, print out an appropriate heading to make the intent of the problem and the tabulation of functional values clear. Finally, draw the graph of each of these functions on the same coordinate system to illustrate this inequality.

4.6.5 Tabulate $\cos^{-1} x$ for $x = [0, .1, 1.0]$. This function may be expressed in terms of ATAN. Special consideration should be given to $\cos^{-1} 0$.

4.6.6 Using identities derived in an introductory study of trigonometric functions, prove each of the following to be identities; complete each statement by deleting values of x for which the equality is not true:

(a) $$\frac{\cot(x/2) - \tan(x/2)}{\cot(x/2) + \tan(x/2)} = \sec x, \qquad x \neq \ldots$$

(b) $$\tan\left(\frac{\pi}{4} + x\right) - \tan\left(\frac{\pi}{4} - x\right) = 2 \tan 2x, \qquad x \neq \ldots$$

(c) $$\tan^{-1} x + \tan^{-1} \frac{1}{2x} = \tan^{-1}\left(2x + \frac{1}{x}\right), \qquad x \neq \ldots$$

the Fourier approximations are

$$S_n(x) = 1 - \frac{2}{\pi} \sum_{k=1}^{n} \left(\frac{\cos k}{k} \sin kx - \frac{\sin k}{k} \cos kx \right), \quad n = 1, 2, 3, \ldots$$

Use identity (2) of Section 4.6 to express $S_n(x)$ as a sum of sines of linear functions of x. Then tabulate $S_3(x)$ and $S_5(x)$ for $x = -5.6 + k(.1)$, where $k = 1, 2, \ldots, 66$ and draw the graphs of $S_3(x)$, $S_5(x)$, and $f(x)$.

4.8 Evaluation of a Polynomial

Given the polynomial function

$$\begin{aligned} p_n(x) &= [a_n, a_{n-1}, \ldots, a_1, a_0] \\ &= a_n x^n + a_{n-1} x^{n-1} + \cdots + a_1 x + a_0 \qquad (1) \\ &= (((a_n x + a_{n-1})x + a_{n-2})x + \cdots + a_1)x + a_0 \qquad (2) \end{aligned}$$

This function can be evaluated for a given value of x either by adding the computed terms of equation (1) or by the successive products and sums of the *nested* form (2). The first method involves $2n - 1$ multiplications and n additions. Computed in its nested form there are n multiplications and n additions. The latter is the more economical procedure.

It can be proved that for any polynomial $P_n(x)$ and any number c, there is a unique polynomial $Q_{n-1}(x)$ and constant R such that

$$P_n(x) = (x - c)Q_{n-1}(x) + R$$

It follows that $P_n(c) = R$. This is the *remainder theorem*. Thus another method of evaluating a polynomial is the division of $P_n(x)$ by $x - c$, the constant remainder being $P_n(c)$.

Consider the algorithm of *synthetic division* to determine the constant remainder associated with divisor $x - c$. Illustrated for $n = 2$, we have

$$\begin{array}{ccc|c} a_2 & a_1 & a_0 & c \\ & a_2 c & (a_2 c + a_1)c & \\ \hline a_2 & a_2 c + a_1 & (a_2 c + a_1)c + a_0 & = R = P_2(c) \end{array}$$

Thus the evaluation of $P_n(c)$ by the remainder theorem and synthetic division is the same algorithm as that of the nested form (2).

4.8.1 About 1225 A.D. Fibonacci (Leonardo of Pisa) attempted an algebraic solution of

$$P_3(x) = x^3 + 2x^2 + 10x - 20 = 0$$

but failed. He then tried to show that a geometric construction of the root (real) by ruler and compass was impossible. These questions reappeared and were answered centuries later.

Note that $P'_3(x) = 3x^2 + 4x + 10 > 0$ for all x. (Why is the derivative everywhere positive?) Thus $P_3(x) = 0$ has but one real root.

If $P_3(x)$ is evaluated for small integer values of x, it can be shown that the single real root lies in the interval (1, 2).

Evaluate $P_3(x)$ for $x = 1. + n(.1)$, where $n = 0, 1, \ldots, 10$, and estimate the value of this root.

4.8.2 Tabulate values of $-\frac{1}{6}(p)(p-1)(p-2)$ for $p = .01n$, where $n = 0, 1, 2, \ldots, 20$. Now compare your results with the tabulation of the Lagrangian coefficient function $A_k^4(p)$, where $k = -1$. (See Table 25.1 in Abramowitz and Stegun under "Tables" in References.)

4.8.3 Repeat example 4.8.1, but after each evaluation, test for the sign of $P_3(x)$. After the first positive value of $P_3(x)$ is found, return to the previous value of x and again evaluate $P_3(x)$ at increments of .01 until $P_3(x)$ becomes positive. Estimate the real root of $P_3(x)$.

4.8.4 By a repeated usage of the device of problem 4.8.3, determine the real root of $P_3(x) = x^3 + 2x^2 + 10x - 20 = 0$ correct to the third decimal place. It will be necessary to determine in this manner the fourth decimal place in order to appropriately round off to three places.

4.8.5 Read in N and $A(1), A(2), \ldots, A(N+1)$, and $B(1), B(2)$, the coefficients of polynomials $P_n(x) = a_1x^n + a_2x^{n-1} + \cdots + a_nx + a_{n+1}$ and $b_1x + b_2$. Print out $C(1), C(2), \ldots, C(N+2)$, the coefficients of the product $(b_1x + b_2)P_r(x)$. Also print out $D(1), D(2), \ldots, D(N)$, the coefficients of the quotient $P_n(x)/(b_1x + b_2)$ and remainder R.

4.9 Factorial Polynomial Function

A special polynomial of degree n whose zeros are $0, 1, 2, \ldots, n-1$ is the factorial polynomial

$$x^{(n)} = x(x-1)(x-2)\cdots(x-n+1)$$

When this polynomial is written in the form

$$x^{(n)} = S_1^n x^n + S_2^n x^{n-1} + \cdots + S_n^n x$$

the coefficients S_i^n, where $i = 1, a, \ldots, n$, are called *Stirling's numbers of the first kind.* For example,

$$\begin{aligned} x^{(1)} &= x &&= x \\ x^{(2)} &= x(x-1) &&= x^2 - x \\ x^{(3)} &= x(x-1)(x-2) &&= x^3 - 3x^2 + 2x \\ x^{(4)} &= x(x-1)(x-2)(x-3) &&= x^4 - 6x^3 + 11x^2 - 6x \end{aligned}$$

n \ i	1	2	3	4	
1	1	0			
2	1	−1	0		
3	1	−3	2	0	
4	1	−6	11	−6	0

(Table heading: S_i^n)

Since $x^{(n+1)} = x^{(n)}(x - n)$, and if each factorial polynomial is expanded in the form of equation (1) of Section 4.8 and coefficients of corresponding powers of x are equated, then

$$S_i^{n+1} = S_i^n - nS_{i-1}^n \qquad (1)$$

This *recurrence relation* states that any element in the triangle of coefficients can be obtained from two elements in the preceding row.

4.9.1 Derive formula (1) of Section 4.9.

4.9.2 Assign variable names to the first column of 1s and the diagonal of 0s in the triangle of Stirling's numbers, with $n \leqq 7$. Using recurrence formula (1) of Section 4.9, assign variable names to the remaining numbers in this array. Print out this triangle of numbers. FORMAT I5 will be sufficient. Programming in integer arithmetic is desirable and feasible since the recurrence relation does not involve division.

4.9.3 Any polynomial can be expressed as a linear combination of factorial polynomials; for example,

$$
\begin{array}{rrrrrr}
P(x) = & x^4 & & -\ 2x^2 & +\ \ x & -\ 7 \\
x^{(4)} = & x^4 & -\ 6x^3 & +\ 11x^2 & -\ \ 6x & \\
\hline
 & & 6x^3 & -\ 13x^2 & +\ \ 7x & -\ 7 \\
6x^{(3)} = & & 6x^3 & -\ 18x^2 & +\ 12x & \\
\hline
 & & & 5x^2 & -\ \ 5x & -\ 7 \\
5x^{(2)} = & & & 5x^2 & -\ \ 5x & \\
\hline
 & & & & & -\ 7
\end{array}
$$

$$P(x) = x^{(4)} + 6x^{(3)} + 5x^{(2)} + 0x^{(1)} - 7x^{(0)}$$

Write a program that will

(a) Store in memory Stirling's numbers, with $n \leq 7$ (problem 4.9.2).

(b) Read in the coefficients of a polynomial of degree 2 and print out the coefficients of the corresponding linear combination of factorial polynomials.

4.9.4 A much more difficult programming problem occurs if we replace 4.9.3(b) by the following: Read in $N \leq 7$ and the coefficients of a polynomial of degree N and print out the coefficients of the corresponding linear combination of factorial polynomials.

4.10 Binomial Coefficient Function

Another special polynomial of degree k whose zeros are at 0, 1, 2, . . . , $k - 1$ is the binomial coefficient function,

$$\binom{S}{0} = 1, \qquad \binom{S}{k} = \frac{S(S-1)(S-2)\cdots(S-k+1)}{k!} = \frac{S^{(k)}}{k!}$$

For positive integral values of $S = n \geq k$, the values of this polynomial $\binom{n}{k}$ are the coefficients of the expanded form of

$$(1 + z)^n = \binom{n}{0} + \binom{n}{1}z + \binom{n}{2}z^2 + \cdots + \binom{n}{n}z^n.$$

Thus we have this name for the polynomial $\binom{S}{k}$. Note that

$$\binom{n}{k} = \frac{n^{(k)}}{k!} = \frac{n!}{k!(n-k)!}$$

Graphs of the polynomials $\binom{S}{k}$, where $k = 0, 1, 2, 3, 4$, are shown in Figure 4.5. For $S = 4$, the values of the sequence of polynomials $\binom{S}{k}$, where $k = 0, 1, 2, 3, 4$, are the coefficients 1, 4, 6, 4, 1 in the expansion of the binomial $(1 + z)^4$.

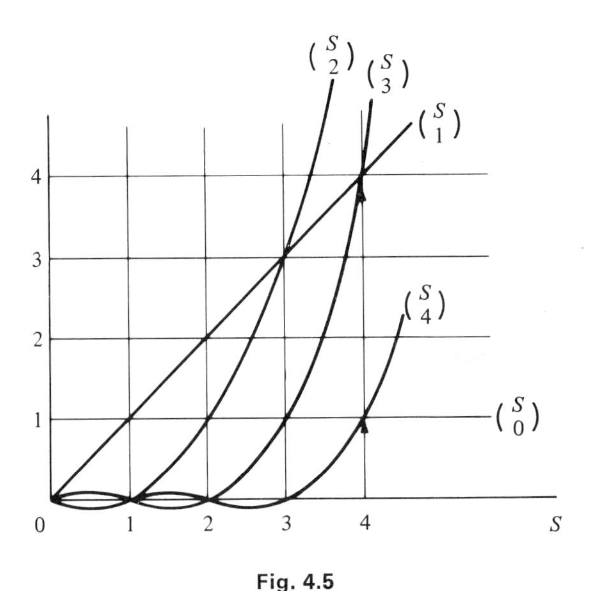

Fig. 4.5

The *binomial coefficients* (values of these polynomials for positive integral values of S) can also be arranged in a triangular array, the Pascal triangle:

	$S \backslash k$	0	1	2	3	4	...
$(1+z)^1 = 1 + z$	1	1	1				
$(1+z)^2 = 1 + 2z + z^2$	2	1	2	1			
$(1+z)^3 = 1 + 3z + 3z^2 + z^3$	3	1	3	3	1		
$(1+z)^4 = 1 + 4z + 6z^2 + 4z^3 + z^4$	4	1	4	6	4	1	
	⋮						

The following relationships obtain among the elements of this triangular array:

$$\binom{S+1}{k+1} = \binom{S}{k+1} + \binom{S}{k} \tag{1}$$

$$\binom{S}{k+1} = \binom{S}{k} \cdot \frac{S-k}{k+1} \tag{2}$$

$$\binom{S+1}{k+1} = \binom{S}{k} \cdot \frac{S+1}{k+1} \tag{3}$$

4.10.1 Derive formulas (1), (2), and (3) in Section 4.10.

4.10.2 Print out Pascal's triangle for $1 \leqq S \leqq 10$. Use integer arithmetic wherever possible.

(a) Assign variable names to the first row and the first diagonal of zeros and use formula (1) of Section 4.10.

(b) Assign variable names to the first column of 1s and use formula (2) of Section 4.10.

(c) Make an appropriate initial assignment of variable names to elements of the triangle and use formula (3) of Section 4.10.

4.11 Polynomial Approximation of a Function

In an interval, a polynomial $P_n(x)$ may be considered an approximation of function $f(x)$ with $R(x)$, the remainder or error:

$$f(x) = P_n(x) + R(x)$$

In Chapter 9, given $n + 1$ functional values $f(x_i)$, where $i = 0, 1, 2, \ldots, n$, we are asked to find $P_n(x)$ so that the error $R(x_i) = 0$, where $i = 0, 1, 2, \ldots, n$. This states that the function and the polynomial are of equal value for $n + 1$ values of x.

Maclaurin's theorem states that $P_n(x)$ is determined under the conditions $R^{(k)}(0) = 0$, where $k = 0, 1, 2, \ldots, n$. In this context $R^{(k)}(x)$ means $d^kR(x)/dx^k$.

Since

$$f^{(k)}(x) = P^{(k)}(x) + R^{(k)}(x), \qquad k = 0, 1, \ldots, n$$

$R^{(k)}(0) = 0$ implies that $f^{(k)}(0) = P^{(k)}(0)$. This states that $P_n(x)$ is constructed so that $f(x)$ and the approximating polynomial are equal in value at $x = 0$ and have their first n corresponding derivatives equal in value at $x = 0$.

4.11.1 As an example of MacLaurin's theorem, in your calculus textbook you will find

$$\sin x = x - \frac{x^3}{3!} + R(x), \qquad |R(x)| < \frac{x^5}{5!}$$

Show that $P_3(x) = x - x^3/3!$ and $\sin x$ are equal in value at $x = 0$ and have common values for their first four derivatives at $x = 0$.

The upper bound that we are given here for the error states that if we use this cubic polynomial in place of $\sin x$ for $0 < x < .1$, the absolute error is less than .0000001.

Tabulate for $x = n(.1)$, where $n = 0, 1, 2, \ldots, 20$:

(a) $\text{Sin}(x)$ as defined by the Fortran subprogram.

(b) $P_3(x) = x - x^3/3!$.

(c) $R(x)$ as defined by (a) and (b).

(d) The upper bound of error given by MacLaurin's theorem, $x^5/5!$.

Draw the graphs of (a) and (b) on the same coordinate system, and also draw the graphs of (c) and (d). Then compare the Fortran subprogram evaluation of $\sin x$ with Table 4.6 in Abramovitz and Stegun.

4.11.2 Apply the instructions of problem 4.11.1 to

$$\log(1 + x) = x - \frac{x^2}{2} + \frac{x^3}{3} + R(x), \qquad |R(x)| < \frac{x^4}{4}, \quad x > 0$$

and compare with Table 4.2 in Abramowitz and Stegun.

4.11.3 Apply the instructions of problem 4.11.1 to

$$e^x = 1 + x + \frac{x^2}{2!} + R(x), \qquad |R(x)| < \frac{e^2}{3!}x^3, \quad 0 < x < 2$$

and compare with Table 4.4 in Abramowitz and Stegun.

4.11.4 Given $e^{-x} = 1. - .9664x + .3536x^2 + R(x)$, where $|R(x)| < 3 \times 10^{-3}$ for $0 \leq x \leq \log 2$. This is listed in the tables in Abramowitz and Stegun as a useful approximation. The upper bound of absolute error (.003) is less than that for the Maclaurin approximation (.05) over this interval.

Write a program to verify this polynomial approximation for $x = k(.01)$, where $k = 0, 1, 2, \ldots, 69$; use the Fortran EXP(−X) as the correct value of e^{-x}.

4.12 Binomial Probability Function

In a course on probability, meaning and use are assigned to the binomial probability function, defined here, and also to the Poisson probability function, defined in Section 4.13. These functions are presented here only because their tabulation is an interesting programming problem thanks to fairly simple recurrence relationships.

For a given positive integer n, and $0 \leqq \theta \leqq 1$, the binomial probability function is

$$b(x) = \binom{n}{x}\theta^x(1-\theta)^{1-x}, \qquad x = 0, 1, 2, \ldots, n.$$

For example, if $n = 3$ and $\theta = \frac{1}{4}$, (Fig. 4.6) then

x	$b(x) = \binom{3}{x}(\frac{1}{4})^x(\frac{3}{4})^{3-x}$
0	$(\frac{3}{4})^3 = \frac{27}{64}$
1	$3(\frac{1}{4})(\frac{3}{4})^2 = \frac{27}{64}$
2	$3(\frac{1}{4})^2(\frac{3}{4}) = \frac{9}{64}$
3	$(\frac{1}{4})^3 = \frac{1}{64}$

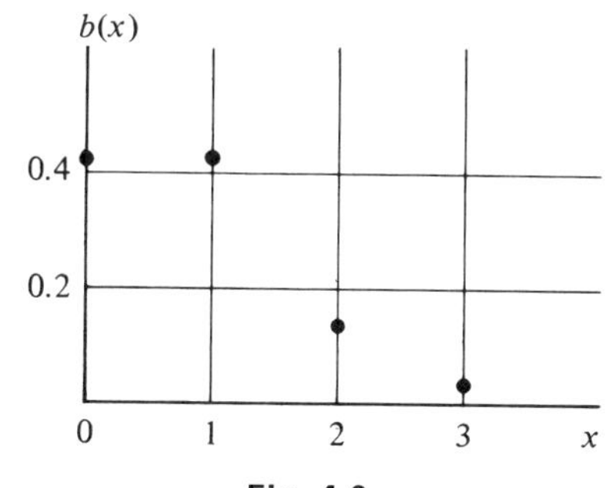

Fig. 4.6

It is interesting to note an important property of this function:

$$\sum_{x=0}^{n} b(x) = \sum_{x=0}^{n} \binom{n}{x} \theta^x (1-\theta)^{n-x} = (1-\theta)^n \sum_{x=0}^{n} \binom{n}{x} \left(\frac{\theta}{1-\theta}\right)^x$$
$$= (1-\theta)^n \left(1 + \frac{\theta}{1-\theta}\right)^n = 1$$

In the above example, add up the values of $b(x)$, where $x = 0, 1, 2, 3$.

As an aid in programming and hand calculation, the following recurrence relation is useful:

$$b(x) = b(x-1) \cdot \frac{n-x+1}{x} \cdot \frac{\theta}{1-\theta}. \tag{1}$$

If an experiment is repeated, and each time a certain outcome has the probability of occurrence θ, then the probability of its occurrence exactly x times in n trials is given by $b(x)$.

4.12.1 Prove the recurrence relationship (1) of Section 4.12.

4.12.2 If $\theta = 2.7\% = .027$, and $n = 50$, tabulate the binomial probability function for $x = 0, 1, 2, \ldots, 20$. Draw the graph of $b(x)$.

What is the value of x having the greatest probability [the largest value of $b(x)$]?

4.12.3 The probability of being dealt a bridge hand with no aces is

$$\theta = \binom{48}{13} \Big/ \binom{52}{13} = .3038$$

Write a program to calculate θ without calculating factorials.

If, in an evening of bridge, 25 hands are dealt, tabulate the probability of having, during that evening, exactly 0, 1, 2, . . . ,25 hands having no aces. What is the most likely number of hands having no aces?

4.13 Poisson Probability Function

For a given positive number μ, the Poisson probability function is defined over the nonnegative integers

$$p(x) = \frac{e^{-\mu}\mu^x}{x!}, \qquad x = 0, 1, 2, \ldots, \quad 0! = 1$$

It is interesting to note an important property of this function:

$$\sum_{x=0}^{\infty} p(x) = e^{-\mu} \sum_{x=0}^{\infty} \frac{\mu^x}{x!}$$

but

$$\begin{aligned} e^{\mu} &= 1 + \mu + \frac{\mu^2}{2!} + \frac{\mu^3}{3!} + \cdots \\ &= e^{-\mu} e^{\mu} \\ &= 1 \end{aligned}$$

As an aid to programming and hand calculation, the following recurrence relation is useful:

$$p(x) = p(x-1)\frac{\mu}{x}, \qquad x = 1, 2, 3, \ldots$$

If an experiment is repeated many times (in Section 4.12, $n \longrightarrow \infty$) and the probability of occurrence of a certain event is the same on each trial, but very small (in Section 4.12, $\theta \approx 0$), then the probability of exactly x occurrences of the event in n trials if $p(x)$, where $\mu = n\theta$.

4.13.1 Compute a table of values of $p(x)$ as follows:

x \ μ	1.	1.5	2.0	2.5	3.0
0					
1					
2					
3					
.					
.					
.					
10					

4.13.2 The cumulative Poisson function is

$$F(x) = \sum_{t=0}^{x} p(t)$$

that is, the chance of exactly x or less occurrences of the event. Tabulate $F(x)$ in a form similar to 4.13.1.

4.13.3 Write a program that reads in a value for μ and prints out the value of x for which $p(x)$ has its greatest value.

4.14 Properties of Conics

A *conic section* may be defined as the locus of points in a plane such that for any one of these points the ratio of its distance to a fixed point (focus) to its distance to a fixed line (directric) is constant (eccentricity e):

If $e < 1$, the locus is an ellipse
If $e = 1$, the locus is a parabola
If $e > 1$, the locus is a hyperbola

4.14.1 Consider the parabola whose equation in a rectangular coordinate system is $y = x^2/4$, whose directrix is $y = -1$, and whose focus is $F(0, 1)$. For the points on the parabola corresponding to $x = 0, 1, 2, 3, 4, 5$, tabulate x, y, PD [the distance from (x, y) to the directrix] PF [the distance from (x, y) to the focus], and $E = PF/PD$.

4.14.2 For the ellipse $x^2/a^2 + y^2/b^2 = 1$, foci are $F_1(c, 0)$, $F_2(-c, 0)$, where $c^2 = a^2 - b^2$ and the corresponding directrices are $x = +a^2/c$ and $x = -a^2/c$. Read in A and $B(A > B)$ and tabulate for $x = n(A/10)$, where $n = 0, 1, 2, \ldots, 10$; $y > 0$, PF_1/PD, and $PF_1 + PF_2$.

4.14.3 For the hyperbola $x^2/a^2 - y^2/b^2 = 1$, foci are at $F_1(c, 0)$ and $F_2(-c, 0)$, where $c^2 = a^2 + b^2$ and the corresponding directrices are $x = a^2/c$, $x = -a^2/c$. Read in A and B and tabulate for $x = A + n[(C - A)/10]$, where $n = 0, 1, 2, \ldots, 10$; $y > 0$, PF_1/PD, and $PF_2 - PF_1$.

4.14.4 Find the point on $y = x^2$ nearest $(1, 0)$. We may approximate this point by computing $l^2 = (x - 1)^2 + y^2$, $y = x^2$ for $x = n(.01)$, where $n = 0, 1, 2, \ldots, 100$, and printing out the coordinates of the point of the parabola associated with the least value of l^2. Show that the analytical solution requires the real root of $2x^3 + x - 1 = 0$.

4.14.5 *Cassini's oval* (Fig. 4.7) is the locus of a point P, the product of whose distances from two fixed points is a constant.

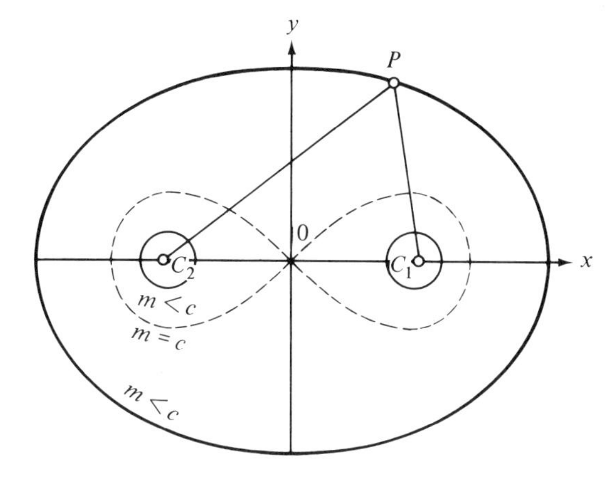

Fig. 4.7

Let the fixed points be $C_1(c, 0)$ and $C_2(-c, 0)$. If $(PC_1)(PC_2) = m^2$, then the polar equation of the locus of P is

$$r^2 = c^2 \cos 2\theta + \sqrt{m^4 - c^4 \sin^2 2\theta}$$

For the special case $m = c$,

$$r^2 = 2c^2 \cos 2\theta$$

the locus is called a *lemniscate.*

For $c = 1$ and $m = 2$ tabulate θ, r, PC_1, PC_2, and $(PC_1)(PC_2)$ for $\theta = 0, \pi/40, 2(\pi/40), \ldots, 10(\pi/40)$.

CONVERGING SEQUENCES

4.15 The Bouncing Ball

A central concern of analysis is the convergence of certain sequences of numbers. We have considered such sequences in defining irrational numbers, the derivative function, the definite integral, and so on.

One sequence that merits considerable attention is the set of sums:

$$S_n(x) = 1 + x + x^2 + \cdots + x^n, \qquad n = 0, 1, 2, \ldots$$

The limit of this sequence for $|x| < 1$ is of simple form and easily demonstrated:

$$(1 - x)S_n(x) = 1 - x^{n+1}$$

$$S_n(x) = \frac{1}{1 - x} - \frac{1}{1 - x}x^{n+1}$$

If we assume the intuitively obvious

$$\lim_{n\to\infty} x^n = 0, \qquad |x| < 1$$

then

$$\lim_{n\to\infty} S_n(x) = \frac{1}{1 - x}, \qquad |x| < 1$$

4.15.1 In the interval [1, 1], carefully sketch the graphs of $1/(1 - x)$ and its successive approximations $S_n(x)$, where $n = 0, 1, 2, 3, 4$.

For $S_{10}(x)$ determine the error $E(x) = 1/(1 - x) - S_{10}(x)$ and the absolute relative error

$$\left| E(x) \middle/ \frac{1}{1 - x} \right|$$

for $-1 < x < 1$.

4.15.2 If a ball is thrown to a height of h feet and rebounds θ times its previous drop $(0 < \theta < 1)$, the distance traveled to its nth impact is

$$D = 2h + 2h\theta + 2h\theta^2 + \cdots + 2h\theta^{n-1}$$

The time required to drop (or rebound) x feet is $\sqrt{2x/g} \approx \sqrt{x}/4$ seconds. The total time elapsed from initial throw to the nth impact is

$$T = 2\frac{\sqrt{h}}{4} + 2\frac{\sqrt{h\theta}}{4}2 + \frac{\sqrt{h\theta^2}}{4} + \cdots + 2\frac{\sqrt{h\theta^{n-1}}}{4}$$

(a) Write closed formulas for D and T.

(b) Find $\lim_{n\to\infty} D$ and $\lim_{n\to\infty} T$.

(c) If a ball is thrown upward to a height of 10 ft and rebounds two thirds of the previous drop, tabulate up to the twentieth impact

(1) The number of the impact.

(2) The height dropped to that impact.

(3) The time required for this drop.
(4) The total distance traveled by the ball to that impact.
(5) The total time elapsed to that impact.

Let the print out be

```
O
              A BALL IS THROWN TO A HEIGHT OF 10 FEET
              AND REBOUNDS 2/3 OF THE PREVIOUS DROP
O
       IMPACT  DROP(FEET)  TIME(SEC)    TOTAL       TOTAL
                                      DISTANCE      TIME
O

O         1              10.      . . .
          2      . . .

O
```

(d) Draw the graphs of TOTAL DISTANCE and TOTAL TIME as a function of the impact number.

4.15.3 → ANNOUNCEMENT—ANNOUNCEMENT ←

On next Feast of Mercury, there will be a footrace between Achilles the speedy and Tortoise the tardy. For this 2000-ft race, Achilles will give his opponent a 1000 ft head start. Although it is known that Achilles' speed is 30 ft/sec, and that of the Tortoise is 1 ft/sec, do not be hasty about predicting the outcome. It is observed that by the time A reaches T's initial position, the latter will have advanced to a farther point, and by the time A reaches this position, T will have moved still farther on. . . . Thus Achilles cannot overtake Tortoise. (Zeno, 430 B.C.)

In the previous problem (part b) we asked: Will the ball stop bouncing? In this problem we ask: Will A catch up to T?

4.16 Important Limits

4.16.1 In the calculus of elementary functions, the following limits are of primary importance:

$$\lim_{\alpha \to 0} \frac{\sin \alpha}{\alpha}, \quad \lim_{\beta \to 0} (1 + \beta)^{1/\beta}$$

It would be appropriate at this point to reflect upon the differentiation of $\sin x$ and $\log x$. If these arguments are not clearly in mind, reference to your calculus textbook is in order.

Tabulate

$$\frac{\sin \frac{1}{2}^k}{\frac{1}{2}^k} \quad \text{and} \quad (1 + \tfrac{1}{2}^k)^{2^k}, \qquad k = 1, 2, \ldots, 10$$

In the same program also tabulate a sequence converging to $\lim_{y \to 0} (1 - y)^{1/y}$. From computed elements of this sequence, guess the value of this limit. Can you justify this guess?

4.16.2 The sequence

$$\frac{1}{n}(e^{1/n} + e^{2/n} + \cdots + e^{n/n}), \qquad n = 1, 2, \ldots$$

converges to a limit. L'Hospital's rule allows us to determine

$$\lim_{n \to \infty} \frac{1}{n}(e^{1/n} + e^{2/n} + \cdots + e^{n/n})$$

(a) Tabulate $1/n \sum_{k=1}^{n} (e^{k/n})$ for $n = 1, 2, 4, 8, \ldots, 32$, and guess the value of this limit.

(b) Find the exact value of this limit.

4.16.3 The limit $\lim_{n \to \infty} n(\sqrt[n]{x} - 1) = \log x$ is a rather unexpected statement. It says that $\log x$ can be approximated by taking square roots.

Tabulate for $x = 2, 4, 6, 8$ the values of $\log x$ as defined by the Fortran subroutine. For each of these values tabulate $n(\sqrt[n]{x} - 1)$ for $n = 2, 4, 8, 16, 32$. A print out might be based on the following:

$x = 2$	$\log 2 =$ (Fortran value)		
n	Approximation	Error	Percent Error
2			
4			
...			
$x = 4$	$\log 4 =$		
...			

What is an apparent inference concerning the variation of "error" with increasing n, x fixed? With increasing x, n fixed? Show these relationships graphically.

4.16.4 Show that $\lim_{x \to 0} (x + e^{x/2})^{2/x} = e^3$.

If you have not yet studied L'Hospital's rule, this may not be a feasible chore. However, we can support this statement by comparing the Fortran value of EXP(3.) with the elements of the sequence

$$(x + e^{x/2})^{2/x}, \qquad x = \frac{1}{2^k}, \quad k = 1, 2, 3, \ldots, 10$$

4.16.5 The sequence of sums

$$\frac{1}{2^2} - \frac{1 \cdot 3}{2^2 \cdot 4^2} + \frac{1 \cdot 3 \cdot 5}{2^2 \cdot 4^2 \cdot 6^2} - \cdots$$

can be shown to converge to some limit. It may not be feasible to find an exact expression for this limit.

Since the terms alternate in sign and decrease in absolute value to zero as n becomes large, any sum

$$S_n = T_1 - T_2 + T_3 - \cdots (-1)^{n+1} T_n$$

differs from the limit

$$L = \lim_{n \to \infty} S_n$$

by less than the absolute value of the next term T_{n+1}:

$$|S_n - L| < T_{n+1}$$

Print out the value of L correct to five decimal places.

4.17 Square Root of a Rational Number

The division algorithm shows that any rational number can be expressed as a repreating decimal; for example,

$\frac{5}{7} = .714285714285714\ldots$	cycle of six digits
$\frac{5}{12} = .416666\ldots$	cycle of one digit

A repeating decimal represents a convergent geometric sequence of rationals whose limit is a rational number; for example,

$$
\begin{aligned}
.3333\ldots &= \frac{3}{10} + \frac{3}{10^2} + \frac{3}{10^3} + \cdots \\
&= \frac{3}{10}\left(1 + \frac{1}{10} + \frac{1}{10}^2 + \cdots\right) \\
&= \frac{3}{10}\left(1\Big/1 - \frac{1}{10}\right) = \frac{1}{3}
\end{aligned}
$$

The square root of a rational number can be expressed as a *repeating continued fraction*; for example,

$$
\begin{aligned}
\sqrt{2} = 1 + \sqrt{2} - 1 &= 1 + \frac{(\sqrt{2} - 1)(\sqrt{2} + 1)}{\sqrt{2} + 1} \\
&= 1 + \frac{1}{\sqrt{2} + 1} \\
&= 1 + \frac{1}{2 + (\sqrt{2} - 1)} \\
&= 1 + \cfrac{1}{2 + \cfrac{1}{2 + \cdots}} \\
&= \langle 1, \bar{2} \rangle
\end{aligned}
$$

This notation represents a continued fraction having 1 in each successive numerator by indicating the integer in each successive denominator. The repeated integers are indicated by the bar.

The sequence of approximations for $\sqrt{2}$ is as follows:

first convergent:	$\langle 1 \rangle = 1.$
second convergent:	$\langle 1, 2 \rangle = 1.5$
third convergent:	$\langle 1, 2, 2 \rangle = 1.4$
fourth convergent:	$\langle 1, 2, 2, 2 \rangle = 1.41666\ldots$

These convergents apparently oscillate about the limit. Refer to the example in 6.4.6.

4.17.1 Given n, write a program to print out the successive continued fraction approximations of $\sqrt{3}$, up to and including the nth.

4.17.2 Approximate $\sqrt{\frac{2}{3}}$ according to the instruction of 4.17.1. Note:

$$\sqrt{\tfrac{2}{3}} = \tfrac{1}{3}\sqrt{6} = \tfrac{1}{3}(2 + \sqrt{6} - 2)$$

↑ greatest integer in $\sqrt{6}$

4.18 Stirling's Formula

There is an important distinction between the concept of *convergence* of a sequence, and sequences that are *asymptotically equal.*

The sequence $\{a_n\}$ converging to limit L implies that $\lim_{n\to\infty} [(a_n - L)] = 0$. If the sequences $\{a_n\}$ and $\{b_n\}$ are such that $\lim_{n\to\infty} (a_n/b_n) = 1$, then $\{a_n\}$ and $\{b_n\}$ are said to be asymptotically equal.

Stirling's formula is a sequence

$$b_n = \sqrt{2\pi}\, n^{n+(1/2)} e^{-n}$$

that is asymptotically equal to n!. (A proof can be found in W. Feller, *Probability Theory and Applications*, New York, John Wiley & Sons, 1968, Vol. 1, Chaps. 2 and 7, and also in G. Chrystal, *Algebra*, New York, Chelsea Publishing Co., 1959 Vol. II, p. 368.)

Another formula similar to Stirling's is

$$c_n = \sqrt{2\pi}\, n^{n+(1/2)} e^{-n+(1/2n)}$$

which also is asymptotically equal to $n!$ It is also shown in Feller that $b_n < n! < c_n$.

4.18.1 Tabulate n, $n!$, $b_n - n!$, $b_n/n!$, c_n, $c_n - n!$, $c_n/n!$ for $n = 1, 2, 3, \ldots, 10$.

4.19 Sequences Converging to π

There are many interesting sequences that converge to the transcendental number π. Examples involving sums, products, and continued fractions are given here. In tabulating these sequences we shall be aware that even if a sequence is known to converge, it may do so with such slowness that it is not useful as an estimator of the limit.

For each of the following, tabulate the first ten convergents and their corresponding errors, using $\pi = 3.141593$:

4.19.1 $$\pi = 4\left[1 - \frac{1}{3} + \frac{1}{5} - \frac{1}{7} + \cdots + (-1)^{n+1}\frac{1}{2n-1} + \cdots\right]$$

4.19.2 $$\pi = \frac{28}{10}\left[1 + \frac{2}{3}\left(\frac{2}{100}\right) + \frac{2.4}{3.5}\left(\frac{2}{100}\right)^2 + \cdots\right] + \frac{30{,}336}{100{,}000}\left[1 + \frac{2}{3}\left(\frac{144}{100{,}000}\right) + \frac{2.4}{3.5}\left(\frac{144}{100{,}000}\right)^2 + \cdots\right]$$

Tabulate this sequence by adding in alternately a term taken from each bracket.

4.19.3 $$\pi = 2\Big/\prod_{n=1}^{\infty}\left[1 - \frac{1}{(2n)^2}\right]$$

4.19.4 $$\pi^2 = 15\Big/\left(1 + \frac{1}{\alpha^2}\right)\left(1 + \frac{1}{\beta^2}\right)\left(1 + \frac{1}{\gamma^2}\right)\cdots$$

where $\alpha, \beta, \gamma, \ldots$ are the primes 2, 3, 5, 7, 11, ...

4.19.5 $$\pi = \cfrac{4}{1 + \cfrac{1^2}{2 + \cfrac{3^2}{2 + \cfrac{5^2}{2 + \cdots}}}}$$

4.19.6 $$\pi = 2\left[1 + \cfrac{1}{1 + \cfrac{1\cdot 2}{1 + \cfrac{2\cdot 3}{1 + \cfrac{3\cdot 4}{1 + \cdots}}}}\right]$$

4.19.7 Given $$\tan^{-1}(x) = \cfrac{x}{1 + \cfrac{x^2}{3 + \cfrac{(2x)^2}{5 + \cfrac{(3x)^2}{7 + \cdots}}}}, \qquad |x| < 1$$

and

$$\tan^{-1} 1 = \tan^{-1}\tfrac{1}{2} + \tan^{-1}\tfrac{1}{3}$$

approximate π printing out 10 convergents tabulated as follows:

convergent $\tan^{-1}(\frac{1}{2})$ $\tan^{-1}(\frac{1}{3})$ $4\tan^{-1}(1)$.

4.19.8 Vieta's fomula is a follows:

$$\frac{\sin x}{x} = \prod_{k=1}^{\infty} \cos\left(\frac{x}{2^k}\right)$$

If $x = \pi/2$, it follows that

$$\pi = 2 \cdot \frac{2}{\sqrt{2}} \cdot \frac{2}{\sqrt{2+\sqrt{2}}} \cdot \frac{2}{\sqrt{2+\sqrt{2+\sqrt{2}}}} \cdots$$

For a proof of Vieta's formula, and a rather unexpected application, refer to M. Kac, *Statistical Independence in Probability Analysis and Number Theory*, New York, John Wiley & Sons, Inc., 1959. This and numerous other trigonometric relationships can be found in E. W. Hobson, *Plane Trigonometry*, New York, Cambridge University Press, 1939.

Using Vieta's formula for $x = \pi/2$, determine a sequence of ten approximations for π.

4.19.9 An interesting repeating, continued fraction converges to the transcendental number e.

$$e = 2 + \cfrac{1}{1 + \cfrac{1}{2 + \cfrac{1}{1 + \cfrac{1}{1 + \cfrac{1}{4 + \cfrac{1}{1 + \cdots}}}}}}$$

$$= \langle 2, \overline{1, 2n, 1} \rangle, \qquad n = 1, 2, 3, \ldots$$

The bar covers the repeated sequence. Write a program to evaluate this continued fraction for $n = 1, 2, 3, \ldots, 10$.

4.20 The Definite Integral

The definite integral of $f(x)$ over interval $[a, b]$ is defined as a limit of a sequence of sums:

1. Partition $[a, b]$ into n subintervals of lengths

$$\Delta x_1, \Delta x_2, \ldots, \Delta x_n$$

2. Let $\xi_1, \xi_2, \ldots, \xi_n$ be values of x each in the interval of the corresponding subscript.

3. Compute the sum $\sum_{i=1}^{n} f(\xi_i)\,\Delta x_i$. If

$$\lim_{\substack{n\to\infty \\ |\Delta x_i|\to 0}} \sum_{i=1}^{n} f(\xi_i)\,\Delta x_i$$

exists,† that number is defined as the definite integral of $f(x)$ over $[a, b]$ and is symbolized by

$$\int_a^b f(x)\,dx$$

4.20.1 For

$$\int_0^1 \frac{dx}{1+x}$$

tabulate

$$\sum_{i=1}^{n} \frac{1}{1+\xi_i}\Delta x_i$$

where ξ_i is always the midpoint of the interval and the lengths of the partitioned intervals are

(a) $\frac{1}{2}, \frac{1}{2}$
(b) $\frac{1}{2}, \frac{1}{4}, \frac{1}{4}$
(c) $\frac{1}{2}, \frac{1}{4}, \frac{1}{8}, \frac{1}{8}$
(d) $\frac{1}{2}, \frac{1}{4}, \frac{1}{8}, \frac{1}{16}, \frac{1}{16}$
(e) $\frac{1}{2}, \frac{1}{4}, \frac{1}{8}, \frac{1}{16}, \frac{1}{32}, \frac{1}{32}$

Since

$$\int_0^1 \frac{dx}{1+x} = \log 2$$

†This expression is independent of the method of partition provided that as $n \longrightarrow \infty$, $\Delta x_i \longrightarrow 0$, where $i = 1, 2, \ldots, n$. It is also independent of the selection of ξ_i in the corresponding interval.

(this is the definition of log 2 in many basic calculus textbooks) and log 2 can be determined by the Fortran subprogram ALOG(2.), write out the corresponding error for each of the sums tabulated. Explain why the error is not decreasing to zero even though $n \longrightarrow \infty$.

4.20.2 For

$$\int_0^1 \frac{dx}{1+x}$$

and the equal interval partition $\Delta x_i = x_i - x_{i-1} = 1/n$, where $i = 1, 2, \ldots, n$, tabulate

$$\sum_{i=1}^{n} \frac{1}{1+\xi i} \Delta x_i$$

for $n = 1, 2, 4, 8, 16$ with

(a) $\xi_i = x_{i-1}$
(b) $\xi_i = x_{i-1} + \frac{1}{4}\Delta x_i$
(c) $\xi_i = x_{i-1} + \frac{1}{2}\Delta x_i$
(d) $\xi_i = x_{i-1} + \frac{3}{4}\Delta x_i$
(e) $\xi_i = x_i$

Is it apparent that the limits of these sequences are independent of the selection ξ_i in each interval?

5

A CONCERN FOR ERROR

5.1 Sources of Error

Our concern for error in this chapter is not based on the inevitable and irritating human mistakes or equipment failures, but on error due to one or more of the following sources:

1. Error in the data upon which the computation is based, and the subsequent propagation of error due to arithmetic operations performed on these data.
2. Error due to the performance of arithmetic operations in the computer in which all numbers are chopped or rounded to a *fixed word length* (a fixed maximum number of significant digits).
3. Error resulting from the use of a difference of two nearly equal numbers where all numbers are chopped or rounded to a fixed word length.
4. Error due to formating output that deletes significant digits.
5. Error inherent in an approximating formula.
6. Error due to the termination of a convergent process.
7. An error in starting some iterative processes, which may result in a convergent sequence—but one having a limit other than the limit that would obtain, if there had been no error in initial steps. A starting error may make some convergent iterative processes unstable and not converge to any limit.

A knowledge of how error may enter a computation certainly does not imply that a simple assessment of the accuracy of a printed result is possible.

If the problem solver has no preconception of what the computed results should be, except perhaps order of magnitude, he has a formidable task in deciding which printed digits are significant. Sometimes this is resolved to his satisfaction by rerunning the problem with slightly different input data or by using approximating formulas with different inherent error; or, he may change the order of arithmetic operations, and thus experimentally build or destroy confidence in the computed results. In most cases a definitive statement on the maximum possible error in a computation is either not possible or economically feasible, let alone a strict accounting of error sources and their relative contribution to the total error.

If the problem solver has some concept of what the results should be, his first reaction to the computer print out might range from complete rejection to acceptance without question. If output is obviously wrong, it's usually back to the drawing board to locate errors in programming or the suitability of the algorithm programmed.

In this text many problems ask us to make appropriate use of approximating formulas and program problems whose result is known beforehand. If, for instance, a print out states that

$$\int_0^1 x^2\,dx = .3333317$$

we might conclude that the algorithm and program are correct and that the obvious error here is a mix of the inherent error of the numerical integration formula and error buildup due to chopping in arithmetic operations. To the student of first-year calculus, the exact answer is $\frac{1}{3}$. To the computing machine with a seven-decimal-digit word length, the exact answer is

$$.3333333$$

For the numerical integration formulas of Chapter 10, it is possible to find an upper bound of the contribution to the error

$$.0000016$$

because of the approximate integration formula. But just how much of the error is due to this and how much is due to other sources of error is a difficult and, in general, not a worthwhile chore to pursue.

In the following sections some examples will be given for the first three of the listed sources of error.

5.2 Error in the Data

In this section "error in the data" does not mean the kind that is due to human blunders—for instance, the key-punching mistake that results in your utility bill being 100 times the usual amount. We are concerned with the validity of numbers resulting from physical measurements and the propagation of errors in these numbers in the progress of a computation.

When a physical measurement involves the counting of the members of a set, the number is an integer. It is an exact physical measurement if the entire set is counted. In cases where the total count is not feasible, such a number may be estimated by inference from counted samples, with a stated probable error. Some data may be defined as exact. A man's hourly wage rate may be exactly the rational number 3.72 dollars per hour or an interest rate of exactly 7.25 per cent.

Decimal approximations of the measure of a physical entity identify an interval containing the actual but unknown measure. The end points of the interval are found by counting standard units. For instance, the weight of an object w might be measured and found to be heavier than 12.5 g and lighter than 13.5 g; that is, its actual weight is in the interval (12.5 g, 13.5 g). This implies that the counted, standard unit is 1 g, and the measure of weight would be recorded as 13 g, the midpoint of the interval. For a more accurate measurement (12.65 g, 12.75 g) the counted units could be .10 g. In the latter case we could write as an exact expression $w = 12.7 + e, |e|_{\max} = .05$. From this measurement the only knowledge we have of the error e are its bounds, $-.05 < e < .05$. The end points of the interval are sometimes written as $w = 12.7 \pm .05$, or more simply $w = 12.7$, implying that 12.7 is a rounded number.

For arithmetic operations on numbers represented by $X = x + e_x$, let an approximate value of X be x, a fixed value (usually the midpoint of an interval of possible values of X), and e_x the error in replacing X by x. If x is the midpoint of an interval (Fig. 5.1) containing X, let the maximum absolute error be $E_x = |e_x|_{\max}$:

Error bounds can be found for the sum, difference, product, and quotient

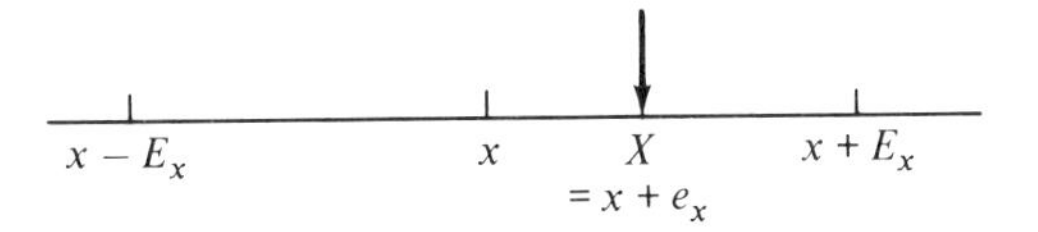

Fig. 5.1

of approximate numbers. For the sum and difference,

$$X \pm Y = x \pm y + (e_x \pm e_y)$$
$$E_{x \pm y} = |e_x \pm e_y|_{\max} = E_x + E_y$$

Thus the maximum absolute error of a sum or difference is the sum of the maximum absolute errors of the terms.

For multiplication and division,

$$XY = (x + e_x)(y + e_y) \approx xy + ye_x + xe_y$$
$$\frac{X}{Y} = \frac{x + e_x}{y + e_y} \approx \frac{x}{y} + \frac{ye_x - xe_y}{y^2}$$

Bounds on relative error are

$$\frac{E_{xy}}{|xy|} = \left|\frac{ye_x + xe_y}{xy}\right|_{\max} = \frac{E_x}{|x|} + \frac{E_y}{|y|}$$
$$\frac{E_{x/y}}{|x/y|} = \left|\frac{(ye_x - xe_y)/y^2}{x/y}\right|_{\max} = \frac{E_x}{|x|} + \frac{E_y}{|y|}$$

Thus the maximum absolute relative error of a product or quotient is the sum of the maximum absolute relative errors of the factors.

For a simple sequence of arithmetic operations it is possible and may be useful to determine an upper bound for error. For a large number of arithmetic operations an upper bound for error may be found but also may not be very informative. In such computations a statistical approach may be used to determine a probable and smaller upper bound for error.

5.2.1 Given $x = 1.3$, a rounded number ($E_x = .05$). Find error bounds on the computed value of $X^2 + 2X + 1$, where coefficients are integers.

$$\begin{array}{rlrl} x^2 &= 1.69 & E_{x^2} &= .13\dagger \\ 2x &= 2.6 & E_{2x} &= .10 \\ 1 &= 1. & E_1 &= .0 \\ \hline x^2 + 2x + 1 &= 5.29 & E_{x^2+2x+1} &= .23 \end{array}$$

†Since $E_{x^2}/x^2 = 2(E_x/x)$, $E_{x^2} = 2xE_x = (2)(1.3)(.05) = .13$

Thus

$$X^2 + 2X + 1\,|_{x=1.3} = 5.29 \pm .23$$

To predict error bounds for the computed value of a differentiable function of a single variable, the differential is a useful approximation of the increment in the function due to a small change in the independent variable:

$$\Delta f \approx df = f'\,dx = f'\,\Delta x$$

5.2.2 Given $x = 1.3$. For a change in this value of x by as much as $\Delta x = .05$, find the corresponding increment in the value of $f(x) = x^2 + 2x + 1$.

$$\begin{aligned}\Delta f &\simeq (2x + 2)\,\Delta x\,|_{x=1.3,\,\Delta x=.05} \\ &= .23\end{aligned}$$

Thus

$$f(1.3) = 5.29 \pm .23$$

To state the computed value of $f(1.3)$, $f(x) = x^2 + 2x + 1$, as $5.29 \pm .23$ is awkward and usually unnecessary. Figure 5.2 shows this interval (1) and the interval for various abbreviated forms. The rounded numbers 5.3 (3) and 5. (4) are plausible abbreviations, the latter ultraconservative. However, the last statement 5.29 (5) implies an error bound that is grossly misleading. A useful rule of thumb is to use no more significant digits in the computed functional value than there are in the least accurate datum.

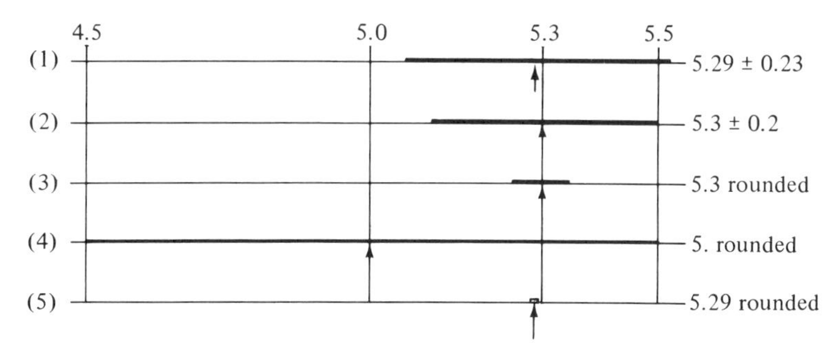

Fig. 5.2

5.2.3 For $f(x) = x^2/(x - 1)$,

(a) Sketch the graph of $f(x)$, $-1 < x < 3$. [Examine f' and f'' and note that since $f(x) = x + 1 + 1/(x - 1)$, $y = x + 1$ is an asymptote.] Refer to Figure 5.3.

(b) For $x = 2.0 - n(.1)$, where $n = 0, 1, 2, \ldots, 9$ and each value of x is considered as a rounded number ($E_x = .05$), print out for each value of x

$$x, \quad f(x), \quad E_{f(x)}, \quad \frac{E_x}{|x|}, \quad \frac{E_{f(x)}}{|f(x)|}$$

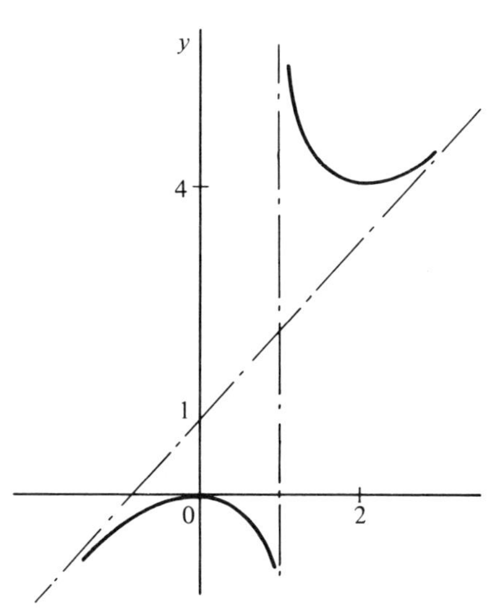

Fig. 5.3

```
      H=.1
      XO=2.
      EX=.05
      DO 10 I=1,10,1
        AI=I-1
        X=XO-AI*H
        F=X*X/(X-1.)
        EF=(1.-1./(X-1.)**2)*EX
      REX=X/ABS(X)
      REF=EF/ABS(F)
   10 WRITE (6,100) X,F,EF,REX,REF
      STOP
  100 FORMAT(5F20.3)
      END
```

x	$\frac{x^2}{x-1}$	E_f	$\frac{E_x}{\lvert x \rvert}$	$\frac{E_f}{\lvert f \rvert}$
2.000	4.000	.000	.025	.000
1.900	4.011	.011	.026	.002
1.800	4.049	.028	.027	.006
1.700	4.128	.052	.029	.012
1.600	4.266	.088	.031	.020
1.500	4.500	.150	.033	.033
1.400	4.899	.262	.035	.053
1.300	5.633	.505	.038	.089
1.200	7.199	1.200	041	.166
1.100	12.100	4.950	.045	.409

If $x = 1.1 \pm .05$ (maximum 4.5 per cent relative error), then $f(x) = 12. \pm 5.$ (maximum 41 per cent relative error).

5.2.4 Consider the given value of x a rounded number. Determine the interval containing the corresponding value of the function. In each case find the maximum absolute relative errors for both x and the function

(a) x^5, $x = 2.$, and $x = 2.00$.
(b) $(x + 1)/(x - 1)$, $x = 2.0$, and $x = 1.1$.
(c) $\log(x - 1)$ and $x = 1.01$.

5.2.5 Sketch the graph of $f(x) = x \log x$ in the interval $0 < x < 3$. Evidence of the behavior of $f(x)$ for x near zero can be supplied by f' and f''.

Consider $x = 1. - n(.1)$, where $n = 0, 1, 2, \ldots, 9$ and each value is a rounded number ($E_x = .05$). Print out for each value of x

$$x, \quad f(x), \quad E_{f(x)}, \quad E_x/|x|, \quad E_{f(x)}/|f(x)|$$

5.2.6 Consider a table of values of $\sin x$; for exact values of $x = n(.001)$, the corresponding values of $\sin x$ are rounded to four decimal places ($E_{\sin x} = 5 \times 10^{-5}$). If we use this table to read $\sin^{-1} x$, what portion of the table gives $\sin^{-1} x$ correct to three decimal places? Write a program to evaluate this largest valid entry for $\sin^{-1} x$.

x		$\sin x$
$\sin^{-1} x$		x
$.xxx$	$.xxxx$	$E = 5. \times 10^{-5}$
$.xxx$	$.xxxx$	

5.3 Errors Due to Fixed Register Arithmetic

A source of computational error that was not fully appreciated before the advent of the digital computer is the computer requirement that all decimals be registered with a fixed maximum number of digits. A fact that further complicates the matter is that most machines convert decimal-number input into an equivalent binary form, and the actual truncation takes place on a binary register with a fixed number of digits. If such a number were converted back to decimal for print out, it would not necessarily be the input decimal truncated to a corresponding number of decimal digits. In this section, for simplification, we will assume that the machine truncates on a decimal register.

A decimal is chopped to t digits when it is replaced by the same decimal with all digits following the tth replaced by zeros. Examples of decimals chopped to the third decimal place are

$$.87381 \longrightarrow .873$$

$$.56695 \longrightarrow .566$$

If a computer retains only t-digit decimals, and numbers are read in with more than t digits, they are chopped to the capacity of the decimal word length. After an arithmetic operation is performed, the result is immediately chopped to t decimal digits. Examples are

exact	three-digit register
$.87381$	$.873$
$+ .56695$	$+ .566$
1.44076	$.143 \times 10^{1}$

$$\begin{array}{r} .87381 \\ \times\ .56695 \\ \hline .4954065795 \end{array} \qquad \begin{array}{r} .873 \\ \times\ .566 \\ \hline .493 \end{array}$$

Each arithmetic operation performed on *integers*, within the limits of word length, is exact (division excluded). But each arithmetic operation performed on *decimal numbers* by a digital computer is, in general, only almost right. Error may not only be introduced at each arithmetic step, but it is propogated throughout the remainder of the computation, as outlined in Section 5.2.

In arithmetic (not the computer's special brand) there is great latitude in the order in which operations can be performed. Basic properties of addition and multiplication are

Commutative:	$A + B = B + A,$	$AB = BA$
Associative:	$A + (B + C) = (A + B) + C,$	$A(BC) = (AB)C$
Distributive:	$A(B + C) = AB + AC$	

In computer decimal arithmetic, because of the necessary restriction of chopping decimals to a finite word length, we can count only on the commutative law of addition; the remainder are only almost correct.

5.3.1 With decimals chopped to three digits find the products (.123)(.789) and (.789)(.123):

$$\begin{array}{r} .123 \\ .789 \\ \hline 1\ 10|7 \\ +\ \ 9\ 84 \\ \hline 109|4 \\ +\ 861 \\ \hline .970 \times 10^{-1} \end{array} \qquad \begin{array}{r} .789 \\ .123 \\ \hline 2\ 3|6|7 \\ +\ 15\ 7|8 \\ \hline 18\ 0 \\ +\ 789 \\ \hline .969 \times 10^{-1} \end{array}$$

In programming, the result of a computation depends on the order in which operations are performed. A striking example of this is a sum of numbers in which there are terms of large and small magnitude. In such a case the obvious way to minimize error is "little ones first."

5.3.2 With decimals chopped to three digits find the sum of .000863 + .00751 + .563:

little first	large first	exact
$.00086\vert 3$	.563	.000863
+ .00751	$+\ .007\vert 51$	.00751
$.008\vert 37$	.570	.563
+ .563	$+\ .000\vert 863$	.571373
.571	.570	

If these terms were rounded numbers, the "exact" sum would (by Section 5.2) have $E \approx .0005$ and a reasonable statement of the sum would be .571, a rounded number.

Part of the mythology that has grown up with the digital computer is its arithmetic infallibility. To be sure, a computer (in proper operating condition) will be able to exactly repeat a decimal computation, but in doing so it will generate the same errors and exactly duplicate their propagation through the computation. In integer arithmetic the computer lives up to its reputation, but in decimal arithmetic it is highly unlikely that it will compute a result exactly.

In arithmetic there is closure for all operations except division by zero. In computer decimal arithmetic, because of the fixed word length, there are restrictions on the *characteristic* of the decimal. For instance, the computer may not be able to store numbers whose absolute value is less than 10^{-75} or greater than 10^{+75}, with the exception of zero. If the computer is asked to perform an operation on two acceptable numbers where the result is too large or too small for the register, the computation is halted because of *overflow*.

5.3.3 If a computer's decimal word length can contain exponents no greater than +75 or less than −75, can you compute and store 70!? Refer to Stirling's formula in Section 4.18.

The novice user of a digital computer should be aware of the presence of error due to fixed-register decimal arithmetic, but should not be overwhelmed. Consider the engineer who is making a stress analysis of a steel structure. His confidence is bolstered by a print out of plausible results. He may experiment by variations in the algorithm or input data or have assurance from previous solutions of similar problems that the first four of seven digits are correct. However, loads on the structure and the modulus of elasticity of the steel may be only known to two digits, and the building-code stress limits are given to only one significant digit. Thus his computer print out with four uncontested digits is adequate assurance to proceed with the building.

5.3.4 Consider the sum $1 + \frac{1}{2} + \frac{1}{2} + \cdots + \frac{1}{8}$. By hand and with two-digit decimal arithmetic,

(a) Find chopped decimal values for each term; add forward, then backward.

(b) Repeat, using rounded approximations.

5.3.5 Program the evaluation of $\sum_{i=1}^{100} 1/n^3$ by adding terms forward, and also by adding backward. Which sum is likely to be nearest the exact sum?

5.3.6 For

$$.000100x + 1.00y = 1.00$$
$$1.00x + 1.00y = 2.00$$

(a) Find the exact decimal expression for the solution.

Using three-digit decimals and chopping,

(b) Solve, eliminating y first.

(c) Solve, eliminating x first.

5.4 Small Differences of Large Numbers

A frequent and disconcerting source of error occurs when the computation requires the difference of nearly equal numbers, for example $2.57 - 2.52 = .05$. If the numbers on the left are rounded numbers, their individual maximum relative error is .2 per cent. The difference .05 then has a maximum relative error of 20 per cent. If the difference is considered as a rounded number, the implied maximum relative error is 10 per cent. One might roughly measure the loss of accuracy due to this arithmetic by observing input data, each with three significant figures, and an output with only one significant figure.

In some computations the arithmetic can be altered to an equivalent computation that avoids the necessity of taking the difference of two nearly equal numbers.

5.4.1 As an example, compute $\sqrt{626.} - \sqrt{625.}$

(a) If we use the algorithm for taking square roots and keep track of the maximum absolute error in both $\sqrt{626.}$ and $\sqrt{625}$, where the maximum absolute error in each radicand is $E = .5$, then

$$\begin{aligned}\sqrt{626.} - \sqrt{625.} &= 25.0 - 25.0 \\ &= .0\end{aligned}$$

(b) If we let $\sqrt{x+1} = \sqrt{x}\,[1 + (1/x)]^{1/2}$ and expand $[1 + (1/x)]^{1/2}$ using the *binomial expansion*,

$$\begin{aligned}\sqrt{x+1} &= \sqrt{x}\left(1 + \frac{1}{x}\right)^{1/2} \\ &= \sqrt{x}\left[1 + \frac{1}{2}\frac{1}{x} + \frac{(\frac{1}{2})(-\frac{1}{2})}{2!}\frac{1}{x^2} + \cdots\right] \\ &= \sqrt{x} + \frac{1}{2}\frac{1}{\sqrt{x}} - \frac{1}{8}\frac{1}{\sqrt{x^3}} + \cdots \\ \sqrt{x+1} - \sqrt{x} &= \frac{1}{2}\frac{1}{\sqrt{x}} - \frac{1}{8}\frac{1}{\sqrt{x^3}} + \cdots\end{aligned}$$

If $x = 625$,

$$\begin{aligned}\sqrt{626} - \sqrt{625} &= \frac{1}{2}\cdot\frac{1}{\sqrt{625}} - \frac{1}{8}\frac{1}{25^3} + \cdots \\ &= .02 - .000008 + \cdots \\ &\simeq .019992\end{aligned}$$

(c) Using the mean-value theorem, $f(x+h) - f(x) = f'(\alpha)h$, where $x < \alpha < x + h$, $f(x) = x^{1/2}$, $h = 1$, and $x = 625$,

$$\begin{aligned}\sqrt{625+1} - \sqrt{625} &= \frac{1}{2\sqrt{\alpha}}, \qquad 625 < \alpha < 626 \\ &\simeq .02\end{aligned}$$

(d) Rationalizing the numerator of $\sqrt{x+1} - \sqrt{x}$,

$$\begin{aligned}\sqrt{x+1} - \sqrt{x} &= \frac{(\sqrt{x+1} - \sqrt{x})(\sqrt{x+1} + \sqrt{x})}{\sqrt{x+1} + \sqrt{x}} \\ &= \frac{1}{\sqrt{x+1} + \sqrt{x}}\end{aligned}$$

If $x = 625$,

$$\frac{1}{\sqrt{626} + \sqrt{625}} \simeq \frac{1}{2\sqrt{625}} = .02$$

(e) Table 3.1 in Abramowitz and Stegun gives the following value:

$$\sqrt{626} - \sqrt{625} = .01999201$$

5.4.2 For large values of x what is a better, alternative form for the computation of

(a) $\sqrt[3]{x+1} - \sqrt[3]{x}$?

(b) $e^{-x} = \cosh x - \sinh x$, given $\cosh x$ and $\sinh x$?

5.4.3 For small values of x what are better alternative forms for the computation of

$$\frac{1 - \cos x}{\sin x}$$

5.4.4 For $x = .9 + n(.01)$, where $n = 0, 1, 2, \ldots, 9$, tabulate both left and right members of

(a) $$\frac{3}{1 - x^3} = \frac{1}{1 - x} + \frac{2 + x}{1 + x + x^2}$$

(b) $$\frac{1 - x^2}{1 - x^3} = \frac{x + 1}{x^2 + x + 1}$$

Considering the right-hand computed value as exact, also tabulate the relative error in the computation of the left-hand member.

5.4.5 For the complex number $z = x + iy$, where $|z| = r = \sqrt{x^2 + y^2}$ and

$$\sqrt{z} = \left(\frac{r + x}{2}\right)^{1/2} + i\left(\frac{r - x}{2}\right)^{1/2}$$

Find the value of $\sqrt{2 + .03i}$

Note that

$$\left(\frac{r - x}{2}\right)^{1/2} = y\Big/2\left(\frac{r + x}{2}\right)^{1/2}$$

Recompute $\sqrt{2 + .03i}$ to avoid the small difference $r - x$.

5.4.6 Let $f(x)$ be a function (Fig. 5.4) that has a second derivative in $[x_0 - a, x_0 + a]$. Let $x_1 - x_0 = x_0 - x_{-1} = h$ and $f(x_i) = y_i$, where $h < a$.

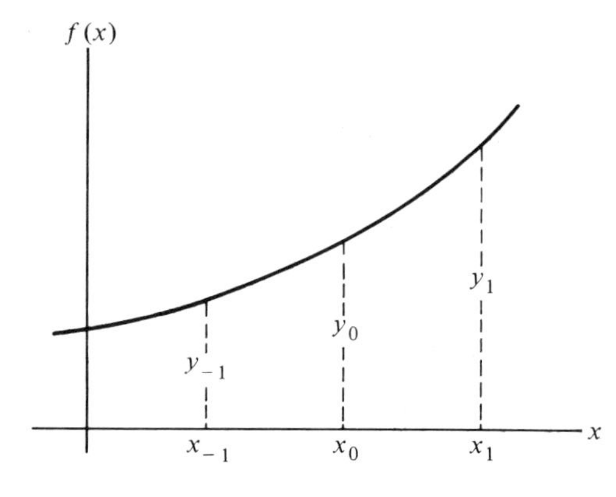

Fig. 5.4

The first derivative is approximated by

$$f'\left(x_0 - \frac{h}{2}\right) \simeq \frac{y_0 - y_{-1}}{h}$$

$$f'\left(x_0 + \frac{h}{2}\right) \simeq \frac{y_1 - y_0}{h}$$

The second derivative may be approximated by

$$f''(x_0) \simeq \frac{f'(x_0 + h/2) - f'(x_0 - h/2)}{h}$$

$$\simeq \frac{y_1 - 2y_0 + y_{-1}}{h^2}$$

Print out the value of sin(1.) as evaluated by the Fortran subroutine.

Using the subroutine to determine $\sin(x \pm h)$, print out a sequence of approximations to

$$\frac{d^2}{dx^2}(\sin x) \text{ of } x = 1. \text{ for } h = .1/5^n, \qquad n = 0, 1, 2, \ldots, 9$$

For each of these approximations print out the relative error. For what values of h is the relative error minimal?

5.4.7 Using the same instructions find a sequence of approximations at $x = 1.$ of $(d^2/dx^2) \tan^{-1} x$ and again find h so that the approximation has least relative error.

5.4.8 Determine a formula to approximate $f^{(4)}(x)$ and compute a sequence of approximations to $(d^4/dx^4)(\cos x)$ at $x = 1$.

6

REAL SOLUTIONS OF $f(x) = 0$

6.1 Initial Estimates

Chapter 4 asked for the evaluation of function $g(x)$ for a given value of x. The inverse of this problem is: For a given value of $g(x)$, find the corresponding value or values of x. This is called "finding the solution of $g(x) = c$," or subtracting c from both members—finding the solution (or roots) of the equation $f(x) = 0$.

In a few instances in our experience we have found the exact value of the roots of an equation. For example, if it is stated that $x^2 - 4x + 3 = 0$, a familiar argument leads to the conclusion that this statement is true if $x = 1$ or $x = 3$.

If it is stated that $x^2 - 4x + 2 = 0$, the same argument implies that $x = 2 - \sqrt{2}$ or $x = 2 + \sqrt{2}$. We have found the solutions of $1 - \sin x = 0$ to be $x = (\pi/2) + n2\pi$, where n is an integer. However, these "exact" solutions may provide cold comfort if we do not know how to find a sequence of decimals that converge to the numbers symbolized by $\sqrt{2}$ and π. If it is stated that $x^2 - \sin x = 0$, "exact" solutions are not obtainable from our brief catalog of methods of solving equations.

The objective of the following problems is to indicate useful methods of estimating real roots of some classes of equations and also to develop iterative arithmetic methods of improving initial estimates of roots. We shall also be concerned with the rate of convergence of the sequence of iterates to the desired value, that is, the arithmetic efficiency of the process.

Iterative methods to determine the roots of an equation require a sufficiently close initial estimate of the root. For a special class of equations, $P_n(x) = 0$, where $P_n(x)$ is a polynomial, there are methods of isolating roots when initial estimates are not known. But these methods (Bernoulli, Graeffe, Quotient-difference) are usually used only to get initial estimates in order to use the more efficient methods discussed in this chapter.

In the exercises of this chapter, we will consider functions that are algebraic expressions of power, trigonometric, and exponential functions and their inverses. The basic properties of these functions have been studied in college freshman and high school courses. If $f(x)$ is not unduly complicated, we may, by "curve sketching," determine the number of real roots of $f(x) = 0$ and estimate their value.

6.1.1 By "curve sketching," estimate the real roots of $x^2 - \sin x = 0$.

(a) Sketch the graph of $y = y_1 + y_2$, where $y_1 = x^2$, and $y_2 = -\sin x$. (See Fig. 6.1.) Estimated values of the x-intercepts of this graph are estimated roots of the equation $x^2 - \sin x = 0$.

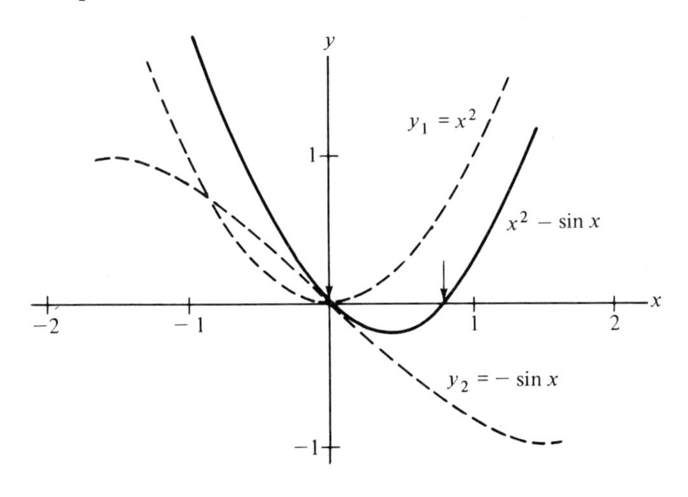

Fig. 6.1

(b) Sketch the graphs of $y_1 = x^2$ and $y_2 \sin x$ (Fig. 6.2) and estimate the abscissa of each point of intersection of these graphs.

It is obvious that there are only two real solutions of $x^2 - \sin x = 0$. A solution $x = 0$ is indicated, and substitution shows it to be an exact solution. The remaining real solution is approximately $x = +.8$.

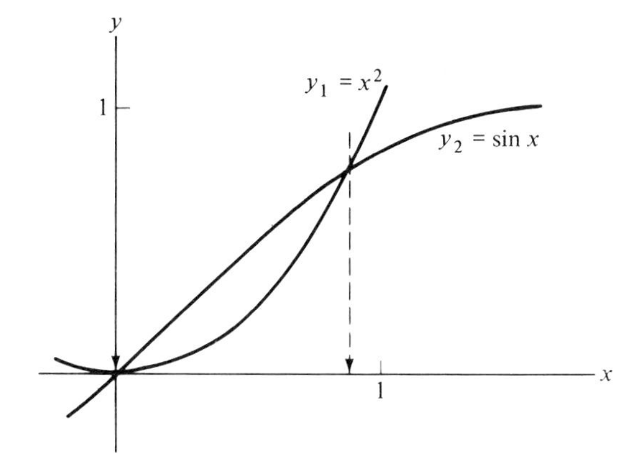

Fig. 6.2

By "curve sketching," estimate the real roots of the following equations. In the remainder of this chapter there are numerical methods to determine closer approximations of roots based on initial estimates. It would be appropriate to use the problems selected for this exercise in subsequent numerical solutions.

6.1.2 $x - \tan x = 0$, where $0 < x < 3\pi/2$.

6.1.3 $x - 2 - \log x = 0$.

6.1.4 $xe^{x^2} - 1 = 0$.

6.1.5 $x^2 = x + 1$; use also the quadratic formula.

6.1.6 $x^7 = x + 1$.

6.1.7 $2x + \sqrt{x} - 1 = 0$; use also the quadratic formula.

6.1.8 $x - 1 - e^{-x^2} = 0$.

6.1.9 $2x - e^{-|x-1|} = 0$.

6.1.10 $\sin x^2 + x - 1 = 0$.

6.1.11 $2 \cos x = \cosh x$.

6.1.12 $x^2 + 1 - \log x = 0$

6.1.13 $\tan x - 1 + x^2 = 0$, where $|x| < \pi/2$.

6.1.14 $\sec^2 x = 2 - x^2$, where $|x| < \pi/2$.

6.1.15 $f' = 0$, if $f = x + \sqrt{x} + 1 + 1/\sqrt{x}$.

6.1.16 $f' = 0$, if $f = x \sin x$, where $0 < x < \pi$.

6.1.17 $f' = 0$, if $f = (\sin x)/x$, where $0 < x < 2\pi$.

6.1.18 $f' = 0$, if $f = e^{-2x} \cos x$, where $|x| < \pi/2$.

In contrast to the preceding two problems, 6.1.18 is easily solved by reference to tables of trigonometric functions.

$$f' = -e^{-2x}(\sin x + 2 \cos x) = 0$$

Since $e^{-2x} \neq 0$,

$$\sin x + 2 \cos x = 0$$

Using identity (1) of Section 4.6,

$$\sqrt{5} \cos (x - x_1) = 0, \qquad x_1 = \tan^{-1} \frac{1}{2}$$

$$x = \tan^{-1} \frac{1}{2} + \frac{\pi}{2} + n \cdot \pi$$

For $n = -1$,

$$\begin{aligned} x &= \tan^{-1} \frac{1}{2} - \frac{\pi}{2} \\ &= -1.107149 \end{aligned}$$

The *mean-value theorem for derivatives* states that for a function $F(x)$, differentiable over interval $[a, b]$, there exists at least one value of x in (a, b) such that

$$F(b) - F(a) = (b - a)F'(u), \qquad a < u < b$$

Find values of u for the following functions and intervals:

6.1.19 $F(x) = xe^{-x}$, [0, 1].

6.1.20 $F(x) = x \sin x$, [0, 1].

6.1.21 $F(x) = 1/(1 + x^2)$, [0, 1].

6.1.22 $F(x) = x^2 \log x$, [1, e].

The *mean-value theorem for integrals* states that for a function $f(x)$, continuous over $[a, b]$, there exists at least one value of x in (a, b) such that

$$\int_a^b f(x)\,dx = (b - a)f(u), \qquad a < u < b$$

Find values of u for the following integrals:

6.1.23 $\int_0^1 (x + e^x)\,dx$.

6.1.24 $\int_0^1 xe^{x^2}\,dx$.

6.1.25 $\int_0^2 (x^2 - \sin x)\,dx$.

6.1.26 $\int_2^3 dx/(x \log x)$.

6.2 Solution by Searching

We may search for a real solution of $f(x) = 0$ by a systematic, repeated evaluation of $f(x)$.

6.2.1 Find the least positive root of $x - \tan x = 0$ (Fig. 6.3). Correct to three decimal places.

Our procedure for this specific problem is based on knowledge of the properties of these elementary functions. We can select a value of x less than this root and know that the corresponding value of f is positive. We shall evaluate $f = x - \tan x$ for $x = 3.5 + n \cdot h$, where $n = 0, 1, 2, \ldots, h = .1$, and observe the sign of f. At a change in sign we shall start from the previous

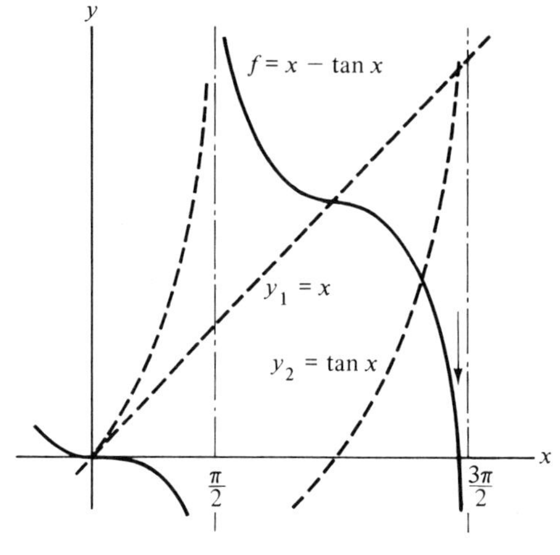

If $f = x - \tan x$
$f' = 1 - \sec^2 x \leqslant 0$

Thus f, decreasing for all x except $x = n.\pi$, is shown in the figure by the solid curve. The least positive root is approximately 4.5.

Fig. 6.3

value of x and repeat the procedure, beginning with the increment $h/10$, until the root is located within a preassigned error. It is rather obvious in this case that our search could miss this root if the initial increment h were too large.

Since a moderately involved strategy is required, it would be useful to first consider a flow chart (Fig. 6.4) and then a program.

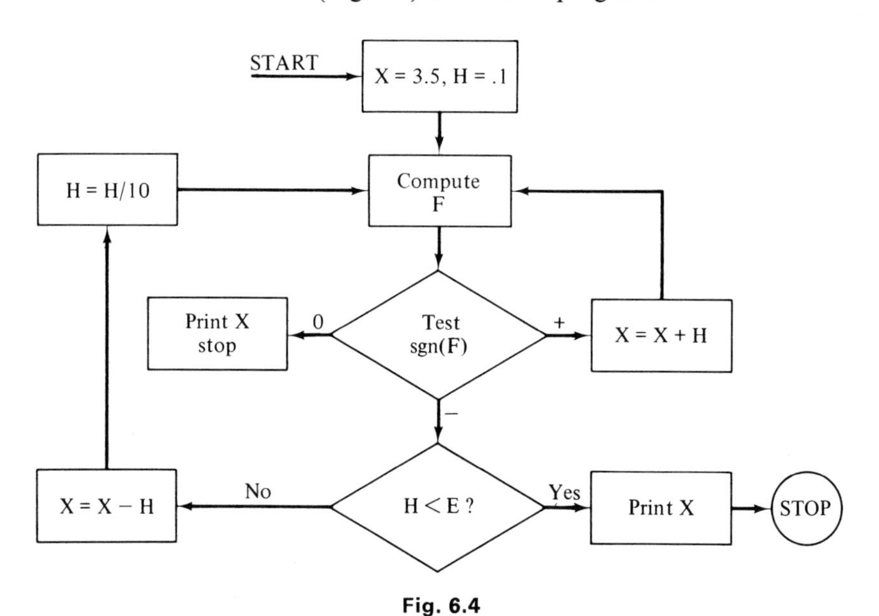

Fig. 6.4

```
      X=3.5
      H=.1
    2 F=X-(SIN(X))/COS(X)
      IF(F)4,3,1
    1 X=X+H
      GO TO 2
    3 WRITE (6,100) X
      STOP
    4 IF(H-.0005)3,3,5
    5 X=X-H
      H=H/10.
      GO TO 2
  100 FORMAT('bbbbTHE LEAST POSITIVE ROOT OF ',
     1'X-TAN(X)=0 IS', F6.3)
      END
```

This program computes four digits to the right of the decimal and chops by printing out only the first three. To print out the rounded approximation to three decimal places one could branch at statement 4 to a rounding routine, as explained in problem 3.18.26.

6.2.2 Select problems from the list of section 6.1 and find roots correct to three decimals by searching.

6.3 Solution by Linear Interpolation

We may determine a real root by combining the searching method with linear interpolation. If $f(x)$ is known to have a single zero in the interval (x_1, x_2), then linear interpolation identifies an interval smaller than $x_2 - x_1$ which contains the root (Fig. 6.5). This procedure is called the *method of false position.*

Let x_3 be the x-intercept of the line containing (x_1, y_1) and (x_2, y_2). Since $(x_3, 0)$ is on the line and $y_1 \neq y_2$,

$$-y_1 = \frac{y_2 - y_1}{x_2 - x_1}(x_3 - x_1)$$

$$x_3 = \frac{x_1 y_2 - x_2 y_1}{y_2 - y_1} = x_1 - \frac{x_2 - x_1}{y_2 - y_1} y_1$$

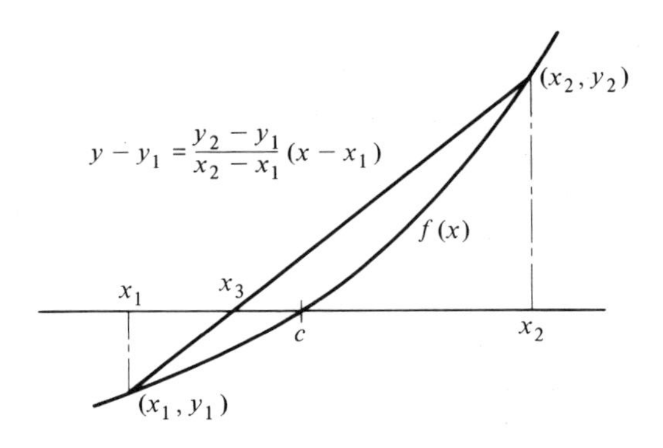

Fig. 6.5

If the criterion for an acceptable approximation is $|f(x)| < E$, a preassigned error bound, a flow chart to determine a corresponding approximate root by a linear interpolation procedure is given in Figure 6.6.

6.3.1 Find the real root of $\sqrt{x} + \log_{10} x = 0$ so that $|f(x)| < .0001$. There is evidently a single real root, and it is in the interval (.1, .5). (See Fig. 6.7.)

A program for solution is as follows:

```
      WRITE (6,100)
      C=.4342945
      X1=.1
      X2=.5
    6 Y1=SQRT(X1)+C*ALOG(X1)
    5 Y2=SQRT(X2)+C*ALOG(X2)
      X3=(X1*Y2-X2*Y1)/(Y2-Y1)
      Y3=SQRT(X3)+C*ALOG(X3)
      AY3=ABS(Y3)
      WRITE (6,101) X3,AY3
      IF(AY3-.0001)1,2,2
    1 STOP
    2 IF(Y1*Y3)3,1,4
    3 X2=X3
      GO TO 5
    4 X1=X3
      GO TO 6
  100 FORMAT('bbbbTHE REAL SOLUTION OF X**.5+LOG(X)=0')
  101 FORMAT('bbbbROOT',F10.7,3X,'WITH ABS(F)=',F10.7)
      END
```

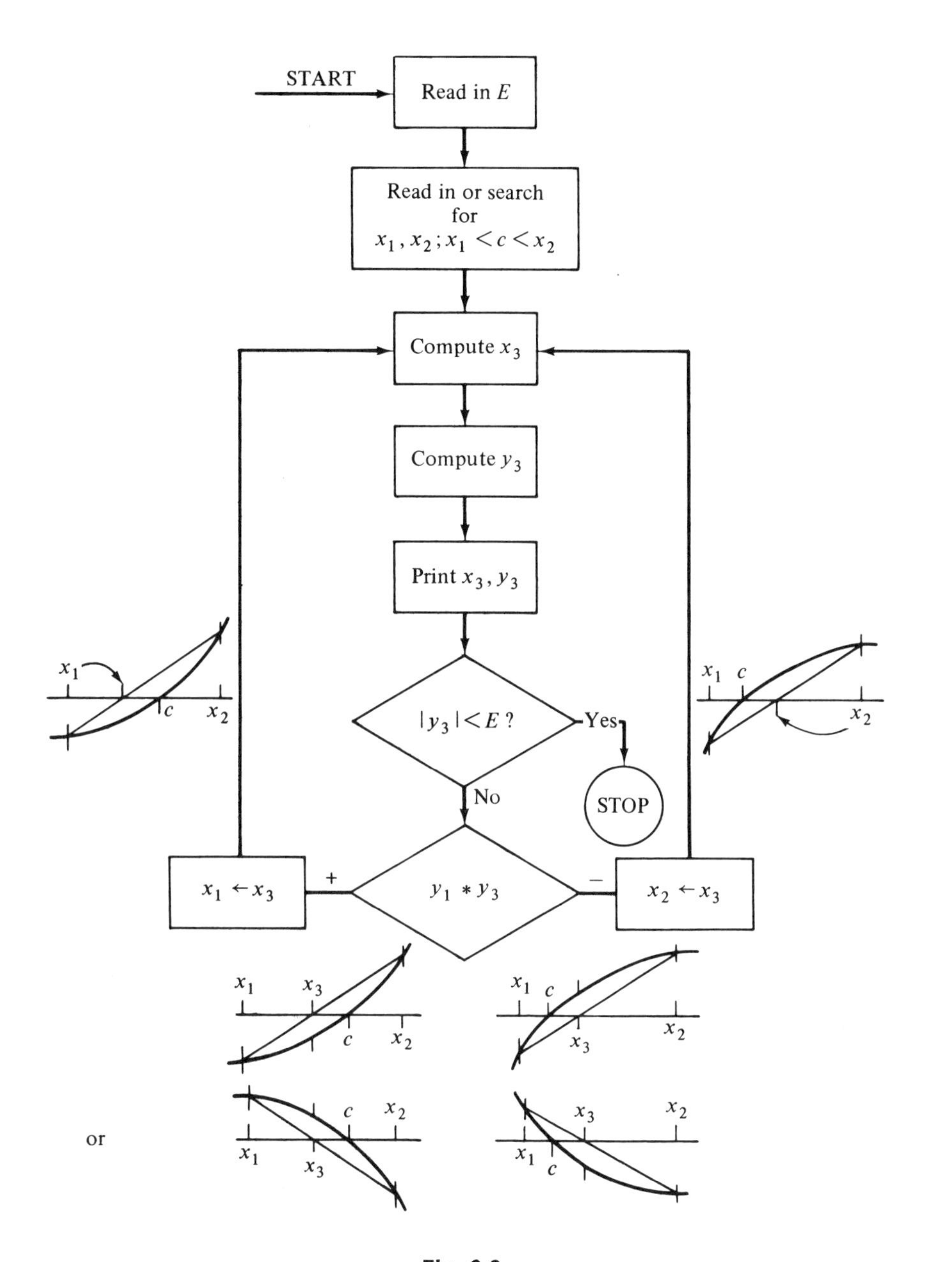

START
Read in E
Read in or search for $x_1, x_2; x_1 < c < x_2$
Compute x_3
Compute y_3
Print x_3, y_3
$|y_3| < E$?
Yes
STOP
No
$y_1 * y_3$
+
−
$x_1 \leftarrow x_3$
$x_2 \leftarrow x_3$
x_1
x_2
x_3
c
or

Fig. 6.6

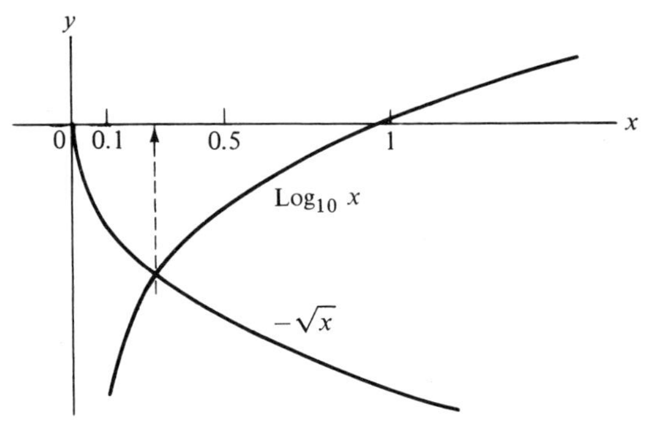

Fig. 6.7

6.3.2 Select any equation of Section 6.1 and find its real roots by linear interpolation such that $|f(x)| < E$, a preassigned bound on the error.

6.4 The Iteration "x=g(x)"

An equation $f(x) = 0$ can be expressed in many ways—in the form $x = g(x)$, for instance, if $f(x) = x - 2 - \log x = 0$, then

$$x = 2 + \log x$$

$$x = e^{x-2}$$

$$x = (e^{2(x-2)} - x)/(x - 1)$$

$$x = \sqrt{2x + x \log x}$$

$$x = x - \frac{x - 2 - \log x}{1 - 1/x}$$

For this equation, the intersections of $y_1 = x$ and $y_2 = 2 + \log x$ shows that there are two real roots and that they are approximately .1 and 3.4.

Consider the sequence defined by $x_1 = 3.4$ and $x_{n+1} = 2 + \log(x_n)$. Using tabulated values of $\log x$,

n	x_n	Change
1	3.400	
2	3.224	176
3	3.170	54
4	3.154	16
5	3.148	6

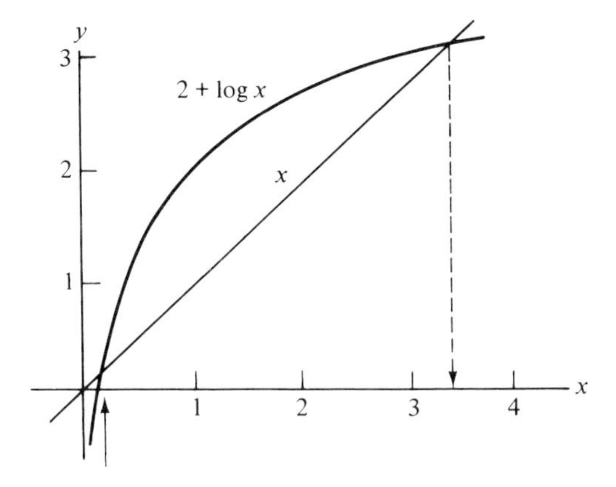

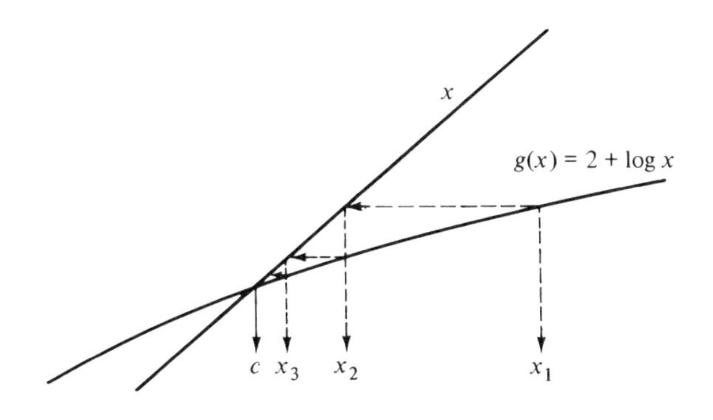

Fig. 6.8

It is evident that this sequence converges, and by examining Figure 6.8 graphically displaying this sequence, we see also that the limit is the larger of the two roots of this equation.

Consider the remaining root and the sequence $x_1 = .2$, $x_{n+1} = 2 + \log(x_n)$.

n	x_n
1	.2
2	.39
3	1.05
4	2.05
.	.
.	.
.	.

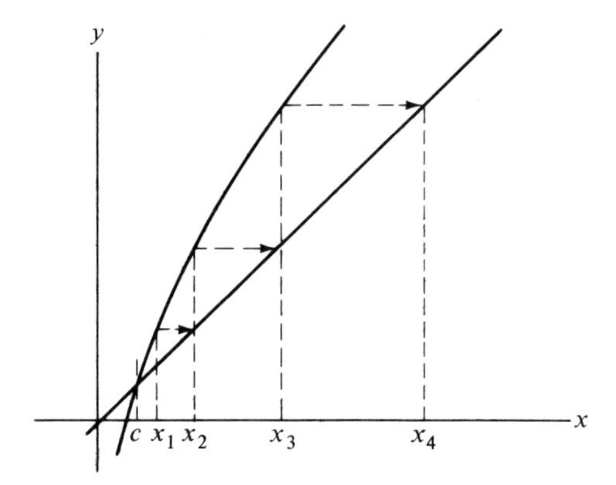

Fig. 6.9

It is evident from the sequence of calculations and Figure 6.9 that the same iterative scheme applied to the smaller root does not result in a sequence converging to that root.

However, the same equation $x - 2 - \log x = 0$ can be expressed in the form $x = e^{x-2}$. Consider the sequence $x_1 = .2$, $x_{n+1} = e^{x_n - 2}$.

n	x_n	change
1	.2	
2	.1653	347
3	.1597	56
4	.1588	9
5	.1586	2

It is evident from the tabulated values and Figure 6.10 that this sequence converges to the smaller of the two roots of $x - 2 - \log x = 0$.

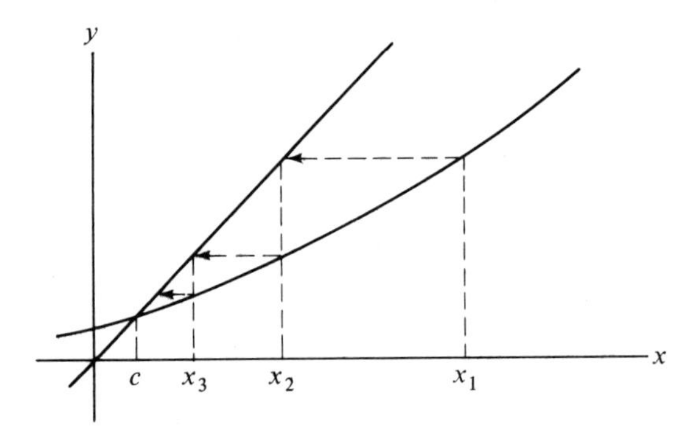

Fig. 6.10

THEOREM If x_1 is in an interval I containing a root $x = c$ of $x = g(x)$, and if $|g'(x)| < 1$ in I, then the sequence defined by $x_{n+1} = g(x_n)$, where $n = 1, 2, \ldots,$ converges to the root, that is, $\lim_{n\to\infty} x_n = c$.

Proof:

$$\begin{aligned} c &= g(c), && \text{since } c \text{ is a root} \\ x_2 &= g(x_1), && \text{definition of } x_2 \text{ in terms of } x_1 \\ x_3 &= g(x_2) \\ &\vdots \\ x_n &= g(x_{n-1}) \end{aligned}$$

Forming differences and applying the mean-value theorem,

$$\begin{aligned} c - x_2 &= g(c) - g(x_1) = (c - x_1)g'(\xi_1), && |c - \xi_1| < |c - x_1| \\ c - x_3 &= g(c) - g(x_2) = (c - x_2)g'(\xi_2), && |c - \xi_2| < |c - x_2| \\ &\vdots \\ c - x_n &= g(c) - g(x_{n-1}) = (c - x_{n-1})g'(\xi_{n-1}), && |c - \xi_{n-1}| < |c - x_{n-1}| \end{aligned}$$

Let $|c - x_i| = e_i$, the absolute error associated with the ith member of the sequence. And let k be the largest value of $|g'(\xi_1)|$. For each value of ξ_1 in interval I, $|g'(\xi_i)| \leqq k < 1$:

$$\begin{aligned} e_2 &= e_1 |g'(\xi_1)| \leqq e_1 k \\ e_3 &= e_2 |g'(\xi_2)| \leqq e_2 k \\ e_n &= e_{n-1} |g'(\xi_{n-1})| \leqq e_{n-1} k \\ e_2 e_3 \cdots e_{n-1} e_n &\leqq e_1 e_2 \cdots e_{n-1} k^{n-1} \\ e_n &\leqq e_1 k^{n-1} \end{aligned}$$

Since $0 < k < 1$, $\lim_{n\to\infty} k^{n-1} = 0$; thus $\lim_{n\to\infty} e_n = 0$. If the absolute error vanishes as n increases, $\lim_{n\to\infty} x_n = c$

It should be noted that in the neighborhood of the root

$$e_n = |g'(\xi_{u-1})| e_{n-1} \approx m e_{n-1}, \qquad m \text{ a constant}$$

This states that the sequence has linear or first-order convergence. If $g'(x)$ near $x = c$ is small, the convergence is rapid. If $|g'(x)|$ is near 1, the convergence is slow. If $-1 < g'(x) < 0$, the sequence oscillates about its limit, and if $0 < g'(x) < 1$, the sequence converges from above or below.

6.4.1 When To Stop

For a nonmachine solution of an equation, "when to stop" may be dictated by the requirements of the problem, the limitations of a slide rule, or human patience. However, a computing machine needs explicit direction.

A customary, and convenient, device is to halt the iteration when two successive approximations differ in absolute value by less than a prescribed amount. For instance,

```
     IF(ABS(X2-X1)-E) n1, n2, n3
n1   STOP
```

It must be remembered that this criterion does not necessarily imply that the error is then less than E. Such a test for stopping the iteration could be met while the error is still undesirably large.

Consider Figure 6.11 which illustrates a slowly, decreasing convergent sequence of the type $x_{n+1} = g(x_n)$ in an interval having $g'(x) \approx +1$.

$$\begin{aligned}|x_n - x_{n-1}| &= |(c + e_n) - (c + e_{n-1})| < E \\ &= |e_n - e_{n-1}| < E\end{aligned}$$

may be met while the error e_n is still large.

By drawing an appropriate figure, illustrate a slowly convergent sequence of the type $x_{n+1} = g(x_n)$ in an interval having $g'(x) \approx -1$. Show through the

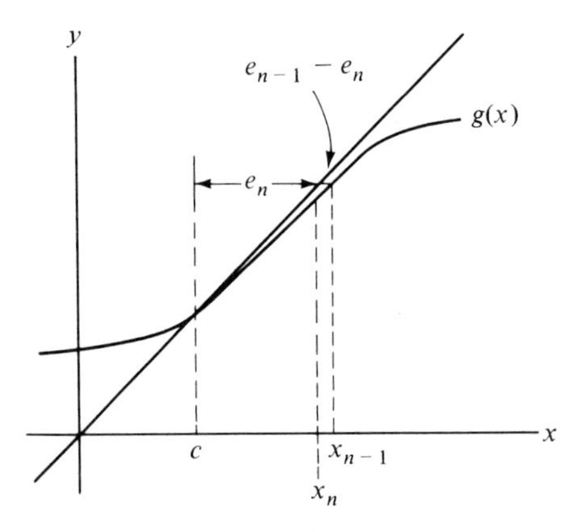

Fig. 6.11

figure that it is possible to have

$$|x_{n+1} - x_{n-1}| < E, \qquad \text{while } e_{n+1} > E$$

In most cases (in particular, in the problems contained in this book), $|g'(x)|$ is not close to 1 in interval I, and the requirement $|x_n - x_{n-1}| < E$ will also imply that $e_n < E$.

It is usually more satisfactory to place the error bound on the approximate relative absolute error ABS((X2−X1)/X2), but again we may be deceived if $|g'(x)| \simeq 1$.

6.4.2 By hand calculation attempt to find the root of

$$x = (.999)(x - 2) + 2$$

The obvious root is 2, but use the iteration $x_{n+1} = g(x_n)$, $x_1 = 3$, and the criterion to stop $|x_n - x_{n-1}| < .01$.

A meaningful test for stopping is the one that we have already used with the example that illustrated solution by interpolation,

```
    IF(ABS(FX)−E) n1, n2, n3
n1  STOP
```

This requires that the approximation of the root be such that the condition $f(x) = 0$ be suitably met.

Another test, requiring some additional programming, identifies an approximation of a root that has a corresponding error less than a prescribed error bound E. For each element of the converging sequence test

$$f(x_n + E) \cdot f(x_n - E)$$

If these two functional values are of opposite sign, then the error

$$e_n = |c - x_n| < E$$

6.4.3 Find the real root of $x^5 - x - 1 = 0$ correct to four significant figures.

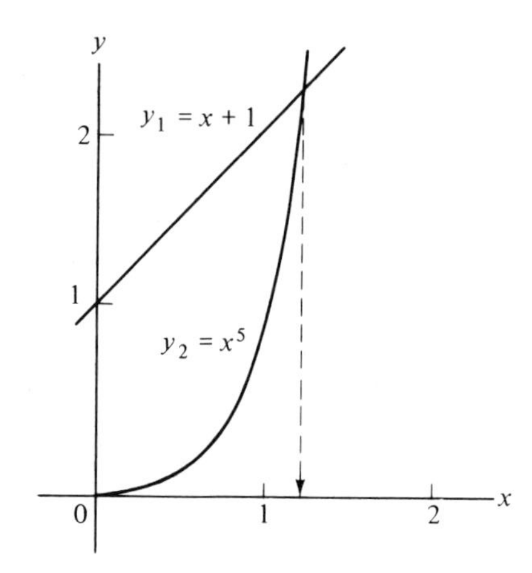

Fig. 6.12

There is one real root at approximately $x = 1.2$ (Fig. 6.12).
If $x = x^5 - 1 = g(x)$,

$$g'(x)|_{1.2} = 5x^4|_{1.2} > 1$$

If $x = (x + 1)^{1/5} = g(x)$, (Fig. 6.13),

$$g'(x)|_{1.2} = \left.\frac{1}{5(x + 1)^{4/5}}\right|_{1.2} < 1$$

Let the sequence be $x_1 = 1.2$, $x_{n+1} = (x_n + 1)^{1/5}$. The rate of convergence, though linear, is rapid since $g'(x)|_{1.2} \approx .1$ Thus each iteration, roughly, provides another correct decimal.

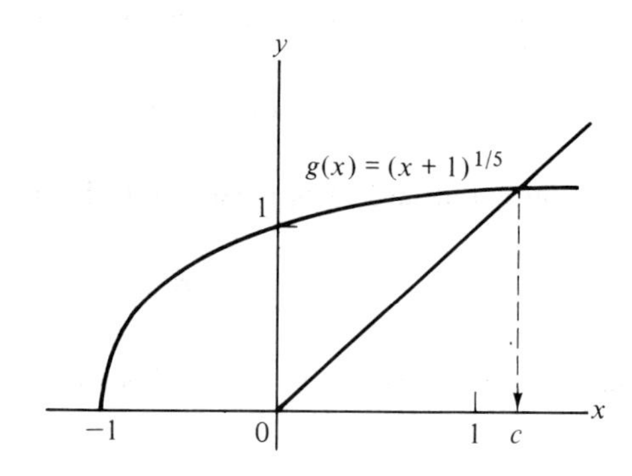

Fig. 6.13

```
                                         Another criterion for stopping
      WRITE (6,100)                    ⎧ XR=X2+.0005
      X1=1.2                           ⎪ XL=X2-.0005
      WRITE (6,101) X1                 ⎪ YR=XR**5-XR-1.
    3 X2=(X1+1.)**.2                   ⎪ YL=XL**5-XL-1.
      WRITE (6,101) X2                 ⎪ IF(YR*YL)1,4,2
      IF(ABS(X1-X2)-.00005)1,2,2⎫     ⎪1 STOP
    1 STOP                      ⎬    ⎨2 X1=X2
    2 X1=X2                     ⎪    ⎪ GO TO 3
      GO TO 3                   ⎭    ⎪4 for this unlikely event find out
                                       ⎪ which is the root, XR or XL, and
                                       ⎩ print out this root
  100 FORMAT('bTHE REAL ROOT OF X**5-X-1=0',//)
  101 FORMAT('b',10X,F6.3)
      END
```

6.4.4 Select problems from the list of Section 6.1 and find real roots by the iteration "$x = g(x)$." In each case use your own, reasonable criterion to STOP.

6.4.5 Print out the first 20 elements of the sequence

$$\sqrt{2+\sqrt{2+\sqrt{2+\sqrt{2+\cdots}}}}$$

Note that if this sequence converges, its limit must be a root of $x = \sqrt{2+x}$ or the positive root of $x^2 - x - 2 = 0$. Prove that the sequence $x_1 = 0$, $x_{n+1} = \sqrt{2 + x_n}$, converges.

In the same manner, consider the sequence

$$\sqrt{6-\sqrt{6-\sqrt{6-\cdots}}}$$

6.4.6 In Section 4.17 continued fractions were evaluated. The first example stated:

$$\sqrt{2} = 1 + \cfrac{1}{2 + \cfrac{1}{2 + \cfrac{1}{2 + \cdots}}} = \langle 1, \bar{2} \rangle$$

If this sequence converges, its limit must be a root of

$$x = 1 + \frac{1}{1+x}$$

Algebraically find the root(s) of this equation. Does the iteration $x_{n+1} = g(x_n) = 1 + (1/1 + x_n)$ converge if x_1 is near $\sqrt{2}$?

Problem 4.17.2 shows that $\sqrt{\frac{2}{3}} = \langle 1, \overline{4, 2} \rangle$. Find a sequence of the form $x_{n+1} = g(x_n)$ that converges to $\sqrt{\frac{2}{3}}$.

6.5 Newton–Raphson Iteration

Basic calculus textbooks usually present an intuitive, graphic concept of the Newton iteration to improve an initial estimate of a root of $f(x) = 0$.

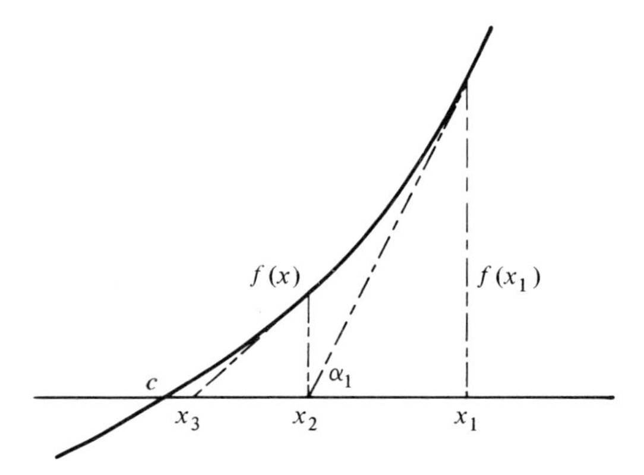

Fig. 6.14

$$\tan \alpha_1 = f'(x_1) = \frac{f(x_1)}{x_1 - x_2}$$

$$x_2 = x_1 - \frac{f(x_1)}{f'(x_1)}$$

With an initial estimate x_1 of the root c, the sequence $x_{n+1} = x_n - f(x_n)/f'(x_n)$, where $n = 1, 2, 3, \ldots$, within the restriction implied by the Figure 6.14, converges to the root c.

6.5.1 Find $\sqrt{13}$ correct to two significant figures.

This number is the positive root of $x^2 - 13 = 0$ (Fig. 6.15). Let $x_1 = 1$ and

$$x_{n+1} = x_n - \frac{x_n^2 - 13}{2x_n} = \frac{1}{2}\left(x_n + \frac{13}{x_n}\right).$$

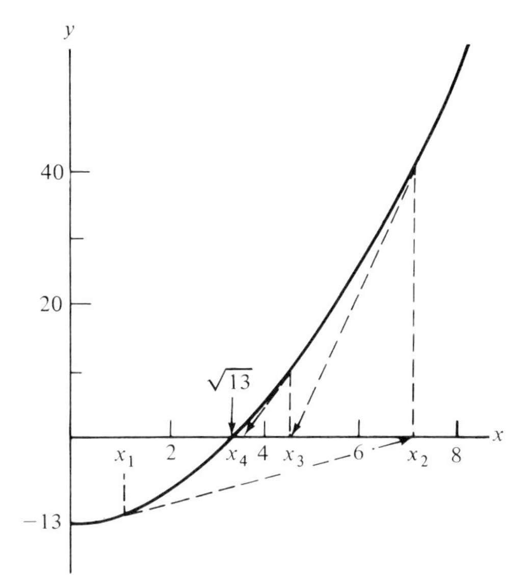

Fig. 6.15

Using a slide rule:

n	x_n	$\frac{1}{2}\left(x_n \frac{13}{x_n}\right)$
1	1	7.
2	7	$\frac{1}{2}(7 + 1.86) = 4.43$
3	4.43	$\frac{1}{2}(4.43 + 2.93) = 3.68$
4	3.68	$\frac{1}{2}(3.68 + 3.53) = 3.605$
5	3.605	$\frac{1}{2}(3.605 + 3.605) = 3.605$

Answer: 3.6

The Newton iteration can be developed from the Taylor expansion of $f(x)$ about x_1, and initial estimate of the root c:

$$f(x) = f(x_1) + f'(x_1)(x - x_1) + \frac{f''(x_1)}{2!}(x - x_1)^2 + \cdots$$

$$f(c) = 0 = f(x_1) + f'(x_1)(c - x_1) + \frac{f''(x_1)}{2!}(c - x_1)^2 + \cdots$$

If $c - x_1$ is small and second-order and higher terms are deleted,

$$0 \simeq f(x_1) + f'(x_1)(c - x_1)$$

An approximate value of c, namely x_2, can be defined by

$$0 = f(x_1) + f'(x_1)(x_2 - x_1)$$

where

$$x_2 = x_1 - \frac{f(x_1)}{f'(x_1)}, \qquad \text{Newton's formula}$$

This suggests that more rapidly converging sequences can be constructed if, for instance, the Taylor series is truncated after the quadratic term.

Again defining x_2, an approximation of root c,

$$0 = f(x_1) + f'(x_1)(x_2 - x_1) + \frac{f''(x_1)}{2!}(x_2 - x_1)^2$$

where

$$x_2 = x_1 - \left.\frac{f' \pm \sqrt{(f')^2 - 2f''f}}{f''}\right|_{x=x_1}$$

This recurrence formula, ordinarily not too useful, could be used to isolate two roots a and b that are almost equal (Fig. 6.16). If the isolated root x_1 of $f'(x) = 0$ is determined, then Newton's quadratic form becomes

$$x_2 = \begin{cases} x_1 - \sqrt{-2f''f}/f'', \\ x_1 + \sqrt{-2f''f}/f'', \end{cases} \qquad f' = 0$$

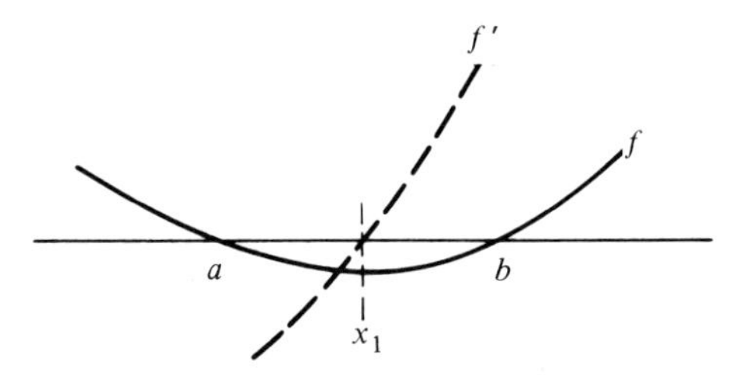

Fig. 6.16

which gives approximately the two almost-equal roots of $f(x) = 0$, each of which can then be improved by Newton's first-order method.

A third view of Newton's iteration is as a special case of the iteration "$x = g(x)$." If $f(x) = 0$, then

$$x = x - \frac{f(x)}{f'(x)} = g(x)$$

and

$$g'(x) = \frac{ff''}{(f')^2}$$

If f'' is continuous, and $f' \neq 0$ in a neighborhood I containing the root c, and since $f(c) = 0$, $g'(c) = 0$. Thus there must be a subinterval in I containing c in which $|g'(x)| < 1$.

This states that under these conditions (f'' continuous, $f' \neq 0$ near $x = c$) Newton's iteration will always converge, if the initial estimates is sufficiently close to the root.

Newton's iteration converges "quadratically," if it converges at all. Since

$$c = g(c)$$

$$x_{n+1} = g(x_n)$$

then

$$x_{n+1} = c = g(x_n) - g(c)$$

If we expand $g(x)$ in a Taylor series about $x = c$, then

$$g(x) = g(c) + g'(c)(x - c) + \frac{g''(c)}{2!}(x - c)^2 + \cdots$$

Substitute, noting that $g'(c) = 0$,

$$x_{n+1} - c = \frac{g''(c)}{2!}(x_n - c)^2 + \frac{g''(c)}{3!}(x_n - c)^3 + \cdots$$

Let the error

$$\begin{aligned} e_{n+1} &= |x_{n+1} - c| \\ &= \frac{|g''(c)|}{2} e_n^2, \qquad \text{deleting higher-ordered terms} \end{aligned}$$

This states that the error in any member of the sequence of approximations is proportional to the *square* of the preceding error. If the error is small, $e_n \approx 10^{-k}$, where $k > 0$, then the next iteration has error $e_{n+1} \approx 10^{-2k}$, doubling the number of significant decimals. For this reason (rapid convergence) Newton's method is frequently used.

If the initial estimate is not close to c, then $g'(x_n)$ may not be near zero, so that Newton's iteration may start out with essentially linear convergence.

When the iterates are close to c, round-off or truncation error may obscure the error associated with the "exact" sequence.

6.5.2 Find the least positive root of $F'(x) = 0$, correct to four decimals, if $F(x) = x^2 \cos x$, $0 < x < \pi/2$.

We note that

$$F'(x) = 2x \cos x - x^2 \sin x$$

if

$$F'(x) = 0, \qquad \tan x = \frac{2}{x}, \quad \text{or } x = 0$$

For Newton's iterative solution of this equation let

$$f(x) = 2 - x \tan x$$
$$f'(x) = -(x \sec^2 x + \tan x)$$

The sketch of the graphs of $2/x$ and $\tan x$ indicate an initial estimate of the solution to be $x_1 = 1.1$, (Fig. 6.17), and the iteration is

$$\begin{aligned} x_{n+1} &= x_n + \frac{2 - x_n \tan x_n}{x_n \sec^2 x_n + \tan x_n} \\ &= \frac{x_n^2 + 2\cos^2 x}{x_n + \cos x_n \sin x_n} \\ &= \frac{x_n^2 + 1 + \cos 2x_n}{x_n + \frac{1}{2}\sin 2x_n} \end{aligned}$$

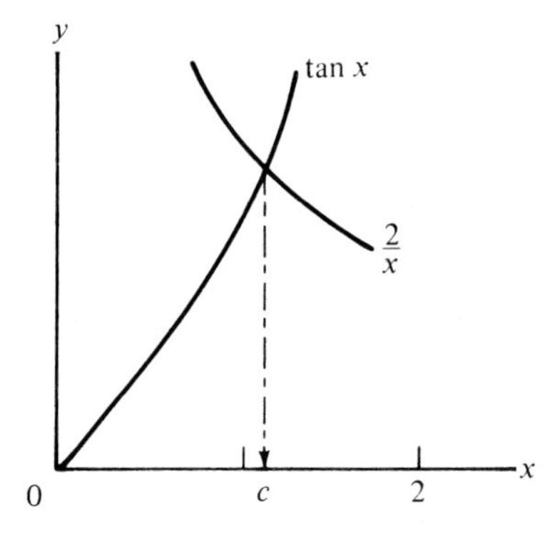

Fig. 6.17

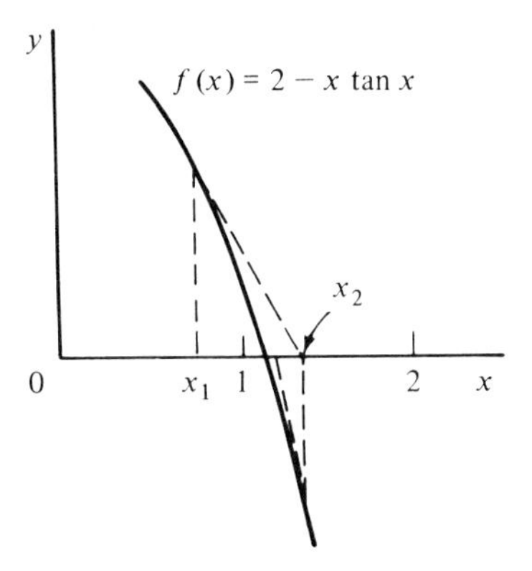

Fig. 6.17 (continued)

A possible program is as follows:

```
      X1=1.1
      WRITE (6,100) X1
    3 X2=(X1*X1+1.+COS(2.*X1))/(X1+.5*SIN(2.*X1))
      WITE (6,100) X2
      IF(ABS(X2-X1)-.00005)1,1,2
    1 STOP
    2 X1=X2
      GO TO 3
  100 FORMAT('b',F7.4)
      END
```

6.5.3 Select problems from the list of Section 6.1 and find real roots by the Newton–Raphson iteration. In each case use your own, reasonable criterion to STOP.

7

REAL SOLUTIONS OF $P_n(x) = 0$

In the problems in Chapter 4 there are various exercises for which it is necessary to evaluate the polynomial function

$$P_n(x) = a_n x^n + a_{n-1} x^{n-1} + \cdots + a_1 x + a_0$$

The inverse is: Given a value of $P_n(x)$, find the corresponding value or values of x. For $n = 1$ and 2, the answer is well known. If

$$P_1(x) = a_1 x + a_0 = b$$

then

$$x = \frac{b - a_0}{a_1}$$

If

$$P_2(x) = a_2 x^2 + a_1 x + a_0 = b$$

then

$$x = \frac{-a_1 \pm \sqrt{a_1^2 - 4a_2(a_0 - b)}}{2a_2}$$

This problem is usually expressed in an equivalent form: Find the value(s) of x for which $P_n(x) = 0$.

A textbook on classical algebra should contain the development of the "exact solutions" for

$$P_3(x) = 0, \quad P_4(x) = 0$$

The "exact solution" is an algorithm expressing the roots of $P_n(x)$ as a finite number of the operations of addition, subtraction, multiplication, division, powers, and roots performed on the coefficients $a_n, a_{n-1}, \ldots, a_0$. A significant portion ofthe history of mathematics has centered about the solution of $P_n(x) = 0$. There follows a list of theorems forming a summary of basic properties of the polynomial function.

7.1 Properties of Polynomials

7.1.1 For $n = 1, 2, 3$, and 4, algebraic formulas are known into which the coefficients of $P_n(x)$ can be substituted to determine all roots of $P_n(x) = 0$. For example, if $ax^2 + bx + c = 0$, there are two roots

$$r_1 = \frac{-b + \sqrt{b^2 - 4ac}}{2a}, \quad r_2 = \frac{-b - \sqrt{b^2 - 4ac}}{2a}$$

7.1.2 For $n \geqq 5$ there are, in general, no algebraic formulas into which coefficients of $P_n(x)$ can be substituted to determine all roots of $P_n(x) = 0$.

7.1.3 There is at least one number r in the complex domain such that

$$P_n(x) = (x - r)Q_{n-1}(x)$$

For example,

$$x^3 - 1 = \left[x - \left(\frac{-1}{2} + \frac{\sqrt{3}}{2}i\right)\right]\left[x^2 + \left(\frac{-1}{2} + \frac{\sqrt{3}}{2}i\right)x - \left(\frac{1}{2} + \frac{\sqrt{3}}{2}i\right)\right]$$

Can you find another value r and write $x^3 - 1$ as a product of a linear and a quadratic factor?

7.1.4 Given $P_n(x)$ and number b, then $P_n(x)$ can be uniquely expressed in the form

$$P_n(x) = (x - b)Q_{n-1}(x) + R, \qquad R \text{ is a constant}$$

It follows that $P_n(b) = R$. For example,

$$x^3 - 1 = (x - 2)(x^2 + 2x + 4) + 7$$
$$2^3 - 1 = 7$$

7.1.5 $P_n(x)$ can be uniquely expressed as a product of n linear factors.

$$P_n(x) = a_n(x - r_1)(x - r_2) \cdots (x - r_n)$$

For example,

$$x^3 - 1 = (x - 1)\left(x + \frac{1}{2} - \frac{\sqrt{3}}{2}i\right)\left(x + \frac{1}{2} + \frac{\sqrt{3}}{2}i\right)$$

The set of numbers $r_1, r_2, \ldots, r_n$ are called the zeros of the function $P_n(x)$ or the roots of the equation $P_n(x) = 0$. If a subset of these zeros of size k are identical, this zero is said to be a multiple or repeated zero of order k. For example,

$$x^4 - 5x^3 + 6x^2 + 4x - 8 = (x - 2)^3(x + 1)$$

The number 2 is a multiple zero of order 3. If multiple zeros are appropriately counted, then $P_n(x)$ has exactly n zeros.

7.1.6 If the coefficients of $P_n(x)$ are real numbers, and if $r = \alpha + i\beta$ is a zero of $P_n(x)$, then the conjugate $\bar{r} = \alpha - i\beta$ is also a zero of $P_n(x)$. For example, since

$$x^3 - 1 = \left[x - \left(-\frac{1}{2} + \frac{\sqrt{3}}{2}i\right)\right]Q_2(x)$$

then

$$x - \left(-\frac{1}{2} - \frac{\sqrt{3}}{2}i\right)$$

is a factor of $Q_2(x)$.

7.1.7 If the coefficients of $P_n(x)$ are real numbers, then the polynomial can be expressed as a product of linear and/or quadratic factors, each with real coefficients. For example,

$$x^3 - 1 = (x - 1)(x^2 + x + 1)$$

Note that

$$[x - (\alpha + i\beta)][x - (\alpha - i\beta)] = x^2 - 2\alpha x + (\alpha^2 + \beta^2)$$

7.1.8

(a) $(d^k/dx^k)P_n(x)$, where $k \leq n$, is a polynomial of degree $n - k$.

(b) $(d^n/dx^n)P_n(x) = a_n n!$.

(c) If $P_n(x) = (x - r)^k Q_{n-k}(x)$, then $P_n(r) = P_n'(r) = \cdots = P_n^{(n-k)}(r) = 0$.

7.1.9 If $P_n(x) = a_n x^n + a_{n-1}x^{n-1} + \cdots + a_1 x + a_0$ has zeros $r_1, r_2, \ldots, r_n$, then

$$\begin{aligned} P_n(x) &= a_n(x - r_1)(x - r_2) \cdots (x - r_n) \\ &= a_n x^n - a_n(r_1 + r_2 + \cdots + r_n)x^{n-1} \\ &\quad + a_n(r_1 r_2 + r_1 r_3 + \cdots + r_{n-1})x^{n-2} + \cdots + (r_1 r_2 \cdots r_n) \end{aligned}$$

Then

$$\sum_{i=1}^{n} r_i = -\frac{a_{n-1}}{a_n}$$

$$\sum_{j=2,i<j}^{n} r_i r_j = +\frac{a_{n-2}}{a_n}$$

$$\sum_{k=3,i<j<k}^{n} r_i r_j r_k = -\frac{a_{n-3}}{a_n}$$

$$\vdots$$

$$r_1 r_2 \cdots r_n = (-)^n \frac{a_o}{a_n}$$

7.2 Methods of Solution

Although the theorem of 7.1.2 states that there is, in general, no finite algebraic formula of coefficients to determine the roots of $P_n(x) = 0$, where $n \geq 5$, the theorems in 7.1.3 and 7.1.4 state that at least one, and at most n, roots do exist.

There is a literature extending over the past three centuries on methods of finding approximate values of the roots $P_n(x) = 0$. Contributions to this problem appear in current journals, largely directed toward more efficient procedures and the problem of isolating roots of nearly equal absolute value.

For the problems of this chapter we will restrict ourselves to finding only real zeros of polynomials. To find complex zeros, the task requires greater insights and numerical effort and is usually included in a course in numerical analysis.

Before presenting two representative algorithms to approximate real roots, there follows a listing, with references and comment, of some of the many devices to determine the roots of polynomial equations.

1. Methods that depend on a known, suitably close initial estimate of the zero.

   (a) Searching, and the linear interpolation methods of Chapter 6. For polynomials these procedures are more cumbersome and require more computer time than (b).

   (b) Newton's method (Chapter 6), which not only converges rapidly, given a suitable initial estimate, but is of simple arithmetic structure.

   (c) Lin's method. An algorithm similar to Newton's requiring less arithmetic per iteration but converging more slowly and under more restrictive conditions. (See Kunz under "Numerical Analysis" in References.)

   (d) Muller's method. A quadratic interpolation technique in current popular use. It can also be used to determine complex zeros. (See Conte under "Numerical Analysis" in References.)

This class of methods will be illustrated by Newton's algorithm.

2. Methods that do not require an initial estimate.

   (a) Bernoulli's method and the quotient-difference scheme. Certain sufficient conditions are known to guarantee the convergence of these iterative procedures. Bernoulli's method identifies the root of largest absolute value, and the quotient-difference scheme finds all roots simultaneously. These methods, in current usage, are being studied in current mathematical literature. (See Henrici under "Numerical Analysis" in References.)

   (b) Lanczos' method of moments (see Lanczos under "Numerical Analysis" in References) and Graeffe's root-squaring method (see Kunz).

The role of these algorithms usually is to isolate roots and supply useful initial estimates for a more rapid computational procedure, for instance Newton's method. This class of algorithms will be illustrated by Graeffe's root-squaring method.

A preference for one method over another depends on whatever prior knowledge is available concerning the roots, the availability of suitable initial estimates, and the experience and preferences of the solver.

7.3 The Newton–Raphson Method

The Newton–Raphson algorithm (Chapter 6) states that if in an interval I, $f(x) = 0$ has a root c, $f(x)$ is differentiable, $f'(x) \neq 0$, and if I contains x_1, a sufficiently close estimate of c, then the iteration

$$x_{k+1} = x_k - \frac{f(x_k)}{f'(x_k)}, \qquad k = 1, 2, 3, \ldots$$

produces a sequence converging to root c quadratically. This algorithm has particular advantages when $f(x)$ is a polynomial. Not only is $P_n(x)$ differentiable everywhere, but both $P'(x)$ and $P_n'(x)$ can be evaluated by a simple arithmetic.

The theorem in 7.1.4, used to determine $P_n(x_k)$, can be extended to determine a similar evaluation of $P_n'(x_k)$. Given $P_n(x)$ and x_k, then there exists $Q_{n-1}(x)$ and constant R such that

$$P_n(x) = (x - x_k)Q_{n-1}(x) + R$$

Thus we have $P_n(x_k) = R$, the "remainder theorem." Differentiating

$$P_n'(x) = (x - x_k)Q_{n-1}'(x) + Q_{n-1}(x)$$

and applying Theorem 7.1.4 to $Q_{n-1}(x)$, we have

$$P_n'(x) = (x - x_k)Q_{n-1}'(x) + (x - x_k)T_{n-2}(x) + S$$

Thus

$$P_n'(x_k) = Q_{n-1}(x_k) = S.$$

Using the synthetic division algorithm, we have

$P_n(x)$:	a_n	a_{n-1}	a_{n-2}	$\cdots$	a_2	a_1	a_0	x_k
$Q_{n-1}(x)$:	b_{n-1}	b_{n-2}	b_{n-3}	$\cdots$	b_1	b_0	$R = P_n(x_k)$	
$T_{n-2}(x)$:	c_{n-2}	c_{n-3}	c_{n-4}	$\cdots$	c_0	$S = P_n'(x)$		

7.3.1 As an example, use a slide rule and find the real root of $x^3 + 2x^2 + 10x - 20 = 0$, with initial estimate $x_1 = 1$.

$$\begin{array}{rrrr|l}
1 & 2 & 10 & -20 & 1. \\
 & 1 & 3 & 13 & \\
\hline
1 & 3 & 13 & -7 = P(1) & \\
 & 1 & 4 & & \\
\hline
1 & 4 & 17 = P'(1) & &
\end{array}
\qquad
\begin{aligned}
x_2 &= 1. - \frac{-7}{17} \\
&= 1.41
\end{aligned}$$

$$\begin{array}{rrrr|l}
1 & 2 & 10 & -20 & 1.41 \\
 & 1.41 & 4.81 & 20.88 & \\
\hline
1 & 3.41 & 14.81 & +.88 = P(x_2) & \\
 & 1.41 & 6.80 & & \\
\hline
1 & 4.82 & 21.61 = P'(x_2) & &
\end{array}
\qquad
\begin{aligned}
x_3 &= 1.41 - \frac{.88}{21.61} \\
&= 1.41 - .041 \\
&= 1.37
\end{aligned}$$

$$\begin{array}{rrrr|l}
1 & 2 & 10 & -20 & 1.37 \\
 & 1.37 & 4.62 & 20.029 & \\
\hline
1 & 3.37 & 4.62 & +.029 = P(x_2) & \\
 & 1.37 & 6.49 & & \\
\hline
1 & 4.74 & 21.11 = P'(x_3) & &
\end{array}
\qquad
\begin{aligned}
x_4 &= 1.37 - \frac{.029}{21.11} \\
&= 1.37 - .0014 \\
&= 1.369
\end{aligned}$$

Therefore, the real root is 1.37 correct to three significant figures.

Since $P_3(x) = (x - 1.37)(x^2 + 3.37x + 14.62)$, the complex roots are

$$-\frac{3.37}{2} \pm \frac{\sqrt{4(14.62) - (3.37)^2}}{2} i$$

$$-1.68 \pm 3.45i$$

7.3.2 Write a program to find all the roots of a cubic equation known to have one real root and two complex roots. Read in the coefficients, the initial estimate of the real root, and an allowable error.

```
READ (5,100) A3,A2,A1,A0,X1,E
WRITE (6,101) A3,A2,A1,A0
B2=A3
```

```
    3 B1=B2*X1+A2
      B0=B1*X1+A1
      R=B0*X1+A0
      C1=B2
      C0=C1*X1+B1
      S=C0*X1+B0
      X2=X1-R/S
      IF(ABS(X2-X1)-E)1,1,2
    2 X1=X2
      GO TO 3
    1 ALPHA=-B1/(2.*B2)
      BETA=SQRT(4.*B2*B0-B1*B1)/(2.*B2)
      WRITE (6,102) X2,ALPHA,BETA
      STOP
      [100,101,102 FORMAT STATEMENTS]
      END
```

Estimate the real roots of the following equations and write programs to find these roots correct to four decimals. Perform at least one of these computations by slide rule to whatever accuracy is possible.

7.3.3 $x^3 - 5x^2 + 2x + 7 = 0.$

7.3.4 $x^3 - x - 1 = 0.$

7.3.5 $x^4 - 2x + .9 = 0.$

7.3.6 $x^{3/2} - (x + 1)^{1/2} = 0.$

7.3.7 $\sqrt{x^3} - \sqrt{2x} - 1 = 0.$

7.3.8 $10x^2 = \sin x$, where $|x| < 1$. Let $\sin x = x - (x^3/3!) + (x^5/5!)$.

7.3.9 Given $F(x) = 1/(1 + x^3)$ in the interval $[0, 1]$. Find all values of x in this interval such that $F(1) - F(0) = (1 - 0)F'(x)$.

7.3.10 Given $f(x) = x^3 + x + 1$, where $a = 0$ and $b = 1$. Find all values of u such that $\int_a^b f(x)\,dx = (b - a)f(u)$. Also find $f(u)$, the average value of $f(x)$ over $[a, b]$.

7.4 Graeffe's Root-Squaring Method

If the roots of a polynomial are widely separated, it is possible to estimate these roots in terms of simple expressions of the coefficients. For example,

$$P_3(x) = (x - 20)(x - 5)(x - 1) = 0$$

$$x^3 - 26x^2 + 125x - 100 = 0$$

$$r_1 = 20 \simeq -\frac{a_2}{a_3} = -\frac{-26}{1} = 26$$

$$r_2 = \ 5 \simeq -\frac{a_1}{a_2} = -\frac{125}{-26} = 4.8$$

$$r_3 = \ 1 \simeq -\frac{a_0}{a_1} = -\frac{-100}{125} = 0.8$$

To demonstrate these approximations, consider the theorem in 7.1.9 at the beginning of this chapter. If

$$\begin{aligned} P_n(x) &= a_n x^n + a_{n-1}x^{n-1} + \cdots + a_1 x + a_0 \\ &= a_n(x - r_1)(x - r_2), \ldots, (x - r_n) \end{aligned}$$

then

$$-\frac{a_{n-1}}{a_n} = r_1 + r_2 + r_3 + \cdots + r_n = r_1\left(1 + \frac{r_2}{r_1} + \cdots + \frac{r_n}{r_1}\right)$$

$$+\frac{a_{n-2}}{a_n} = r_1 r_2 + r_1 r_3 + \cdots + r_{n-1}r_n = r_1 r_2\left(1 + \frac{r_3}{r_2} + \cdots + \frac{r_{n-1}r_n}{r_1 r_2}\right)$$

$$\begin{aligned} -\frac{a_{n-3}}{a_n} &= r_1 r_2 r_3 + r_1 r_2 r_4 + \cdots + r_{n-2}r_{n-1}r_n \\ &= r_1 r_2 r_3\left(1 + \frac{r_4}{r_3} + \cdots + \frac{r_{n-2}r_{n-1}r_n}{r_1 r_2 r_3}\right) \end{aligned}$$

$$\vdots$$

If $r_1, r_2, r_3, \ldots,$ are all real and distinct and ordered $|r_1| > |r_2| > |r_3| >$

$\cdots > |r_n|$ and widely separated, that is, $r_2/r_1 \simeq 0$, $r_3/r_2 \simeq 0, \ldots,$

$$-\frac{a_{n-1}}{a_n} \simeq r_1 \qquad\qquad r_1 \simeq -\frac{a_{n-1}}{a_n}$$

$$+\frac{a_{n-2}}{a_n} \simeq r_1 r_2 \qquad \text{and} \qquad r_2 \simeq -\frac{a_{n-2}}{a_{n-1}}$$

$$-\frac{a_{n-3}}{a_n} \simeq r_1 r_2 r_3 \qquad\qquad r_3 \simeq -\frac{a_{n-3}}{a_{n-2}}$$

$$\vdots$$

$$r_n \simeq -\frac{a_0}{a_1}$$

Graeffe's root-squaring method to approximate the roots of $P_n(x) = 0$ depends upon an algorithm to construct the coefficients of a second polynomial $Q_n(y)$ whose zeros are the square of those of $P_n(x)$. If the roots of $P_n(x) = 0$ are distinct and $|r_i| > |r_{i+1}|$, then

$$0 < \left|\frac{r_{i+1}}{r_i}\right| < 1$$

The corresponding roots of $Q_n(y)$ are r_i^2 and r_{i+1}^2. Since

$$0 < \frac{r_{i+1}^2}{r_i^2} < \left|\frac{r_{i+1}}{r_i}\right| < 1$$

the roots of $Q_n(y) = 0$ are more widely separated than those of $P_n(x) = 0$.

An algorithm that would construct the coefficients of

$$Q_n(y) = b_n y^n + b_{n-1} y^{n-1} + \cdots + b_1 y + b_0$$

from the coefficients of $P_n(x)$ can be repeatedly used to form a sequence of polynomials. The roots of any polynomial in the sequence are the squares of the corresponding roots of the preceding one. Each polynomial then has roots that are more widely separated than those of the previous polynomial. If the root-squaring process is repeated k times,

$$r_i^{2^k} \simeq -\frac{b_{n-i}}{b_{n-i+1}}$$

or

$$|r_i| \simeq \sqrt[2^k]{-\frac{b_{n-i}}{b_{n-i+1}}}, \qquad i = 1, 2, 3, \ldots, n$$

may be satisfactory approximations of the roots of $P_n(x) = 0$.

Given the polynomial $[a_n, a_{n-1}, \ldots, a_1, a_0]$, an algorithm to construct the polynomial $[b_n, b_{n-1}, \ldots, b_1, b_0]$ whose zeros are the squares of the given polynomial is indicated in diagram form for $n = 5$. Its extension to higher- or lower-ordered polynomials is evident.

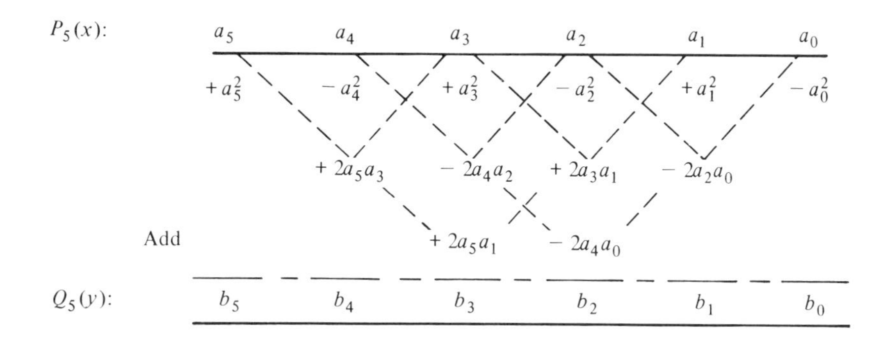

For instance, $b_2 = -a_2^2 + 2a_3a_1 - 2a_4a_0$.

To verify this algorithm for $n = 2$,

$$P_2(x) = a_2x^2 + a_1x + a_0 \qquad Q_2(y) = b_2y^2 + b_1y + b_0$$

$$= a_2(x - r_1)(x - r_2) \qquad = b_2(y - r_1^2)(y - r_2^2)$$

$$r_1 + r_2 = -\frac{a_1}{a_2} \qquad r_1^2 + r_2^2 = -\frac{b_1}{b_2}$$

$$r_1r_2 = \frac{a_0}{a_2} \qquad r_1^2r_2^2 = \frac{b_0}{b_2}$$

Solving for coefficients b_2, b_1, and b_0, each in terms of coefficients a_2, a_1, and a_0,

$$b_2 = a_2^2, \quad b_1 = -a_1^2 + 2a_0a_2, \quad b_0 = a_0^2$$

7.4.1 Given that the roots of $P_3(x) = x^3 - 26x^2 + 125x - 100$ are real and distinct. Using the root-squaring algorithm twice and a slide rule, estimate the roots of this equation.

				Approximate $\lvert r_i \rvert$		
$P_n(x)$: 1	-26	125	-100	26	4.8	.8
1	-676 $+250$	1.56×10^4 $-.52$	-10^4			
$Q_n(y)$: 1	-426	$+1.04 \times 10^4$	-10^4	$\sqrt{426}$ $= 20.4$	$\sqrt{24.4}$ $= 4.9$	$\sqrt{.96}$ $= .98$
1	-18.1×10^4 $+2.08$	$+1.08 \times 10^8$ $-.805$	-10^8			
1	-16.02×10^4	$+.995 \times 10^8$	-10^8	$\sqrt[4]{160{,}200}$ $= 20.0$	$\sqrt[4]{625}$ $= 5.0$	$\sqrt[4]{1.00}$ $= 1.0$

Note that this iteration approximates only the absolute value of the roots. Substitution into $P_n(x)$ is needed to determine the proper sign.

7.4.2

(a) Write the fourth-degree polynomial whose roots are $-2, 1, 3, 4$.

(b) Given the coefficients a_4, a_3, a_2, a_1, a_0, and using the root-squaring algorithm twice, estimate these roots with the aid of a slide rule.

(c) Write a program that will print out the absolute value of these roots correct to the second decimal. (After each step, test to determine if the difference between successive root estimates is less than .005 for *all* roots.)

(d) Amend this program to print out the roots correct to the second decimal.

7.4.3 Write a program to print out the roots of $x^3 - x^2 - 22x + 39 = 0$ correct to four decimal places. All roots of this cubic are real.

8

LINEAR SYSTEMS OF EQUATIONS

The problem of this chapter is: Given coefficients $[a_{ij}]$ and (c_i), find a set of numbers (x_j) such that

$$\sum_{j=1}^{n} a_{ij}x_j = c_i, \qquad i = 1, 2, \ldots, n$$

For $n = 3$ these linear equations in unknowns (x_1, x_2, x_3) are

$$\begin{aligned}
a_{11}x_1 + a_{12}x_2 + a_{13}x_3 &= c_1 \\
a_{21}x_1 + a_{22}x_2 + a_{23}x_3 &= c_2 \\
a_{31}x_1 + a_{32}x_2 + a_{33}x_3 &= c_3
\end{aligned}$$

A student's initial experience with this problem usually involves solvable systems of only two or three unknowns, and coefficients that are small integers, for example

$$\begin{aligned}
2x - 3y &= 1 \\
x + 2y &= 4
\end{aligned}$$

By the argument *elimination of unknowns* we construct the equivalent system (a system of equations having the same solution):

$$x = 2$$
$$y = 1$$

The moves made in this game are called row operations. To continue this illustration involving two equations, let the equations be represented by

$$l_1(x, y) = 2x - 3y - 1 = 0$$

and

$$l_2(x, y) = x + 2y - 4 = 0$$

If a number pair (x, y) is such that

$$l_1(x, y) = 0 \qquad \text{and} \qquad l_2(x, y) = 0$$

then this number pair also satisfies the linear equation

$$c_1 l_1(x, y) + c_2 l_2(x, y) = 0$$

for any choice of c_1 and c_2, except both zero.

The equation $c_1 l_1 + c_2 l_2 = 0$ can be one of a system equivalent to a given system containing $l_1 = 0$ and $l_2 = 0$. In the above example

$$\tfrac{2}{7}(2x - 3y - 1) + \tfrac{3}{7}(x + 2y - 4) = 0$$

reduces to

$$x - 2 = 0$$

Thus an appropriate choice for the multipliers in this case is

$$c_1 = \tfrac{2}{7}, \quad c_2 = \tfrac{3}{7}$$

Another choice of multipliers can produce the equivalent equation,

$$y - 1 = 0$$

The success of this procedure applied to problems of studied simplicity depends on the student's ingenuity to invent an efficient elimination procedure for the particular system. For instance, the equivalence of

$$\begin{aligned} x + y + 3z &= 10 \\ 3x - y - z &= -8 \\ x + 3y - z &= 2 \end{aligned} \qquad \text{and} \qquad \begin{aligned} x &= -1 \\ y &= 2 \\ z &= 3 \end{aligned}$$

can be demonstrated readily by several simple strategies.

If a linear system is presented for solution where there are, for example, ten equations in ten unknowns and all coefficients are decimals with several significant figures, then ad hoc strategies are obviously not appropriate. For large-order systems a general, systematic, and efficient procedure must be employed.

8.1 Methods of Solution

There are two types of solution methods, direct and iterative.

A *direct* solution of a solvable system leads to the exact solution through a finite number of algebraic operations. The familiar elimination method is of this type. A systematic elimination of unknowns usually leading to an equivalent "triangular" system as an intermediate step is called the *Gaussian elimination method;* for example,

$$\begin{aligned} x + y + 3z &= 10 \\ 3x - y - z &= -8 \\ x + 3y - z &= 2 \end{aligned} \qquad \begin{aligned} x + y + 3z &= 10 \\ y + \tfrac{5}{2}z &= \tfrac{19}{2} \\ z &= 3 \end{aligned} \qquad \begin{aligned} x &= -1 \\ y &= 2 \\ z &= 3 \end{aligned}$$

Successive elimination of unknowns by row operations in the given system produces the equivalent triangular system. Back substitution leads to the equivalent system explicitly stating the solution. We will illustrate the direct solution with the Gauss-Crout method.

An *iterative* method involves an initial estimate of the solution and an algorithm to compute a sequence of approximations that converge to the solution. For example,

$$\begin{aligned} 3x - y - z &= -8 \\ x + 3y - z &= 2 \\ x + y + 3z &= 10 \end{aligned}$$

Let an initial estimate be $x^{(0)} = y^{(0)} = z^{(0)} = 0$ and let subsequent estimates be found by

$$\begin{aligned} x^{(k+1)} &= \tfrac{1}{3}[-8 \qquad\qquad\quad + y^{(k)} + z^{(k)}] \\ y^{(k+1)} &= \tfrac{1}{3}[\ \ 2 - x^{(k)} \qquad\quad + z^{(k)}], \qquad k = 0, 1, 2, \ldots \\ z^{(k+1)} &= \tfrac{1}{3}[\ 10 - x^{(k)} - y^{(k)} \qquad\quad] \end{aligned}$$

A few iterations can easily be obtained with the aid of a slide rule:

k	x	y	z
0	.0	.0	.0
1	−2.667	+.667	+3.333
2	−1.333	+2.667	+4.000
3	−.444	+2.444	+2.889
4	−.889	+1.778	+3.037

In this illustration it is at least plausible that the sequence converges to the solution. We will illustrate iterative procedures with the Gauss–Seidel method and state sufficient conditions for convergence.

8.2 Solvability

A sufficient condition that a linear system have a solution is that the determinant of the coefficients of the left members not vanish.

$$D = |a_{ij}| \neq 0$$

This statement can be readily verified for the case $n = 2$:

$$\begin{aligned} a_{11}x_1 + a_{12}x_2 &= c_1 \\ a_{21}x_1 + a_{22}x_2 &= c_2 \end{aligned}$$

If we eliminate unknowns,

$$x_1 = \frac{c_1 a_{22} - a_{12}c_2}{a_{11}a_{22} - a_{12}a_{21}}, \quad x_2 = \frac{a_{11}c_2 - c_1 a_{21}}{a_{11}a_{22} - a_{12}a_{21}}$$

Let

$$D = \begin{vmatrix} a_{11} & a_{12} \\ a_{21} & a_{22} \end{vmatrix} = a_{11}a_{22} - a_{12}a_{21}$$

Then

$$x_1 = \frac{1}{D}\begin{vmatrix} c_1 & a_{12} \\ c_2 & a_{22} \end{vmatrix}, \quad x_2 = \frac{1}{D}\begin{vmatrix} a_{11} & c_1 \\ a_{21} & c_2 \end{vmatrix}$$

Thus if the divisor D is not zero, there is a solution. The solution expressed in terms of determinants is called Cramer's rule.

Consider the graphs of

$$l_1 \equiv a_{11}x_1 + a_{12}x_2 - c_1 = 0, \quad x_2 = -\frac{a_{11}}{a_{12}}x_1 + \frac{c_1}{a_{12}}$$

$$l_2 \equiv a_{21}x_1 + a_{22}x_2 - c_2 = 0, \quad x_2 = -\frac{a_{21}}{a_{22}}x_1 + \frac{c_2}{a_{22}}$$

In the first graph (Fig. 8.1) the lines intersect (the system has a unique solution) if the slopes are not equal.

$$-\frac{a_{11}}{a_{12}} \neq -\frac{a_{21}}{a_{22}}$$

$$a_{11}a_{22} - a_{12}a_{21} = D \neq 0$$

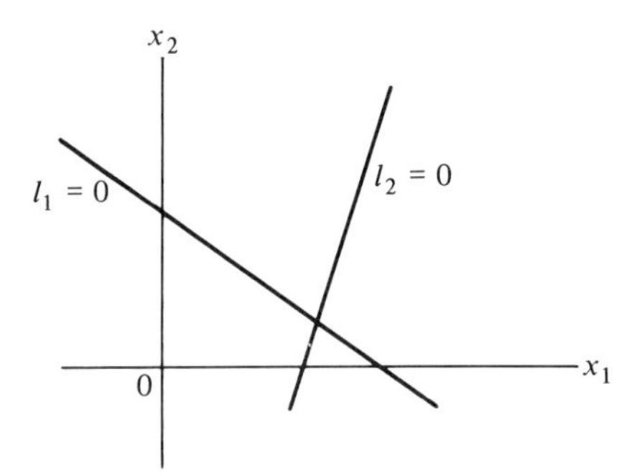

Fig. 8.1

$D \neq 0$ is a sufficient condition for the system to have a solution, and only one solution.

In the second graph (Fig. 8.2) the lines will not intersect (the system is inconsistent), if slopes are equal and x_2 intercepts are not equal.

$$D = 0 \quad \text{or} \quad -\frac{a_{11}}{a_{12}} = -\frac{a_{21}}{a_{22}}, \qquad \text{equal slopes}$$

$$\begin{vmatrix} c_1 & a_{12} \\ c_2 & a_{22} \end{vmatrix} \neq 0, \qquad \text{unequal } x_2 \text{ intercepts}$$

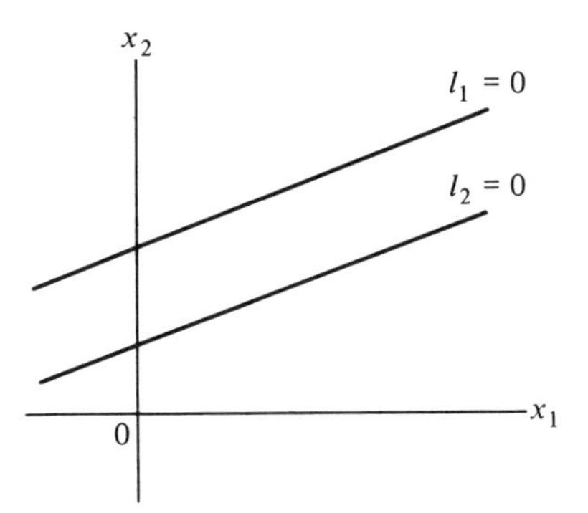

Fig. 8.2

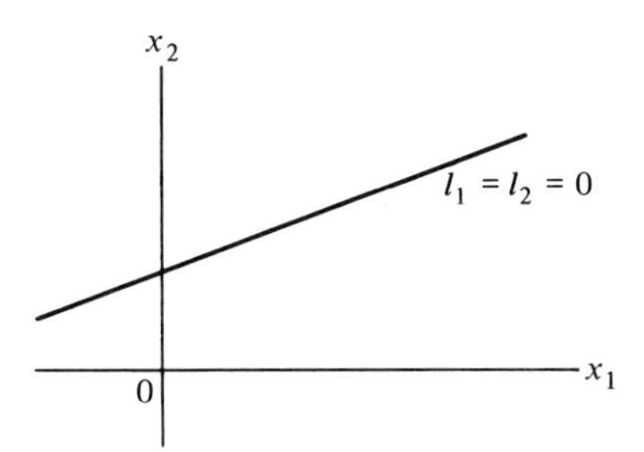

Fig. 8.3

In the third graph (Fig. 8.3) the lines will coincide (the system is consistent, having an infinite set of solutions) if

$$D = 0 \quad \text{and} \quad \begin{vmatrix} c_1 & a_{12} \\ c_2 & a_{22} \end{vmatrix} = 0$$

or if corresponding coefficients are proportional: $a_{11}/a_{21} = a_{12}/a_{22} = c_1/c_2$.

For the homogeneous system in which $c_1 = c_2 = 0$, the following two cases are of interest.

In the fourth graph (Fig. 8.4) the lines intersect at the origin (the homogeneous system has the trivial solution $x_1 = x_2 = 0$) if $D \neq 0$.
In the fifth graph (Fig. 8.5) the lines contain the origin and also coincide (the homogeneous system has nontrivial solutions) if $D = 0$ or if corresponding coefficients are proportional: $a_{11}/a_{21} = a_{12}/a_{22}$.

We can consider, without verification at this time, extensions of these five cases to linear systems of higher order.

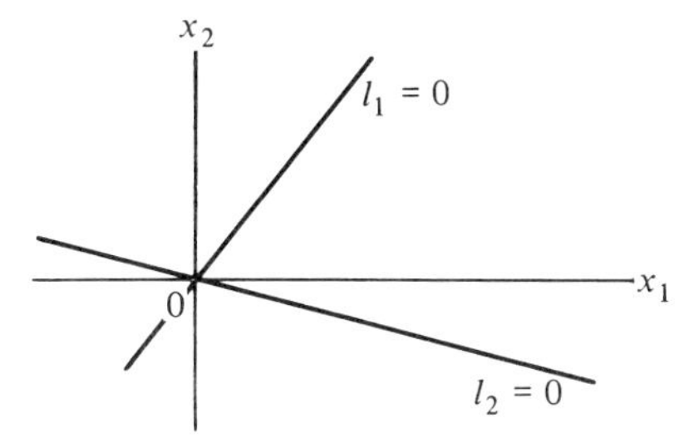

Fig. 8.4

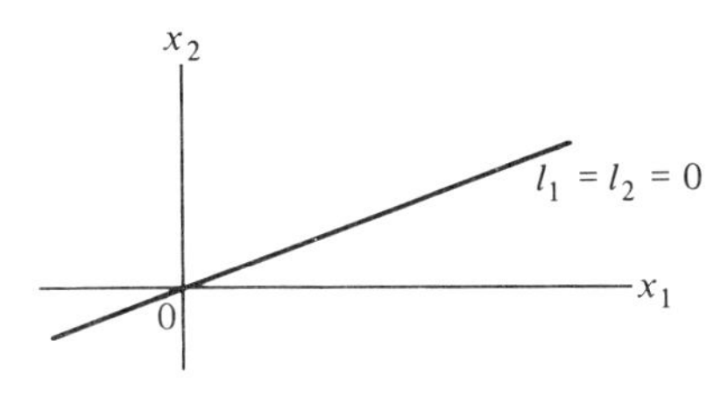

Fig. 8.5

Given a nonhomogeneous system for which $D \neq 0$, the extension of the first case tells us that a unique solution exists. With a direct solution method, a finite number of arithmetic operations produces this solution. It is somewhat disconcerting to realize that in practice we might never be able to compute this solution. The fine-print proviso is that a direct solution is guaranteed only if the result of each arithmetic operation is exactly known. In a computing machine (or a slide rule), arithmetic is performed with a fixed number of significant figures. Thus the result of an arithmetic operation on two numbers usually is a chopping of the exact result with a corresponding introduction of error.

The coefficients of a linear system having $D \neq 0$ may be such that the buildup of error eventually destroys all significant figures at some point in the computation. The property of the coefficients $A = [a_{ij}]$ that makes the computed solution have a large probable error, or not possible to find, is a large value of the *condition of* A, Cond(A). The definition of this property and its evaluation is a subject for in numerical analysis. In such a course arithmetic strategies would be examined to minimize the buildup of error for those systems for which Cond(A) is large. It is important to note that even if D is near zero, Cond(A) could be small and a computed solution guaranteed. We will consider for solution in this book linear systems for which solvability is not an issue, and also avoid the problem of assessing probable error in the computed result.

8.2.1 Write a program to read in by rows the 2×3 array $[a_{ij}]$, and for the linear system

$$a_{11}x_1 + a_{12}x_2 = a_{13}$$
$$a_{21}x_1 + a_{22}x_2 = a_{23}$$

print out, correctly, either

NO SOLUTION

or

X1 = ---------, X2 = ---------

or

X1 = T , X2 = ---------T

8.3 The Gauss–Crout Solution

Consider, for $n = 3$, the successive elimination of unknowns (Gaussian elimination) to obtain an equivalent triangular system:

$$\begin{array}{l|c|c|c} a_{11}x_1 + a_{12}x_2 + a_{13}x_3 = a_{14} & \dfrac{1}{a_{11}} & -a_{21} & -a_{31} \\ a_{21}x_1 + a_{22}x_2 + a_{23}x_3 = a_{24} & & +1 & \\ a_{31}x_1 + a_{32}x_2 + a_{33}x_3 = a_{34} & & & +1 \end{array} \tag{1}$$

The notation at the right of this system indicates the sequence of row operations to find an equivalent system having x_1 eliminated from all but the first equation. For example, the given system (1) is transformed into the equivalent system (2) by the following row operations:

$$l_4 = \frac{1}{a_{11}} l_1 \qquad = 0$$
$$l_5 = -a_{21}l_4 + (1)l_2 = 0$$
$$l_6 = -a_{31}l_4 + (1)l_3 = 0$$

$$\begin{array}{l|c|c|l} x_1 + b_{12}x_2 + b_{13}x_3 = b_{14} & & & b_{1j} = \dfrac{a_{1j}}{a_{11}}, \quad j = 2, 3, 4 \\ c_{22}x_2 + c_{23}x_3 = c_{24} & \dfrac{1}{c_{22}} & -c_{32} & c_{2j} = a_{2j} - a_{21}b_{1j} \\ c_{32}x_2 + c_{33}x_3 = c_{34} & & +1 & c_{3j} = a_{3j} - a_{31}b_{1j} \end{array} \tag{2}$$

$$\begin{aligned} x_1 + b_{12}x_2 + b_{13}x_3 &= b_{24} \\ x_2 + b_{23}x_3 &= b_{24} \qquad & b_{2j} &= \frac{c_{2j}}{c_{22}}, \quad j = 3, 4 \\ d_{33}x_3 &= d_{34} \left| \frac{1}{d_{33}} \right. & d_{3j} &= c_{3j} - c_{32}b_{2j} \end{aligned} \tag{3}$$

$$\begin{aligned} x_1 + b_{12}x_2 + b_{13}x_3 &= b_{14} \\ x_2 + b_{23}x_3 &= b_{24} \\ x_3 &= b_{34} \qquad & b_{34} &= \frac{d_{34}}{d_{33}} \end{aligned} \tag{4}$$

Back substitution leads to the solution

$$\begin{aligned} x_1 \qquad &= b_{14} - b_{13}x_3 - b_{12}x_2 \\ x_2 \quad &= b_{24} - b_{23}x_3 \\ x_3 &= b_{34} \end{aligned} \tag{5}$$

The Crout method is an improved bookkeeping device to perform these same arithmetic operations. This device is helpful in either a hand computation or in programming Gaussian elimination for a machine.

Let A be the matrix of the left members augmented by the right members as an added column:

$$A = \begin{bmatrix} a_{11} & a_{12} & \cdots & a_{1n} & a_{1,n+1} \\ a_{21} & a_{22} & \cdots & a_{2n} & a_{2,n+1} \\ \cdot & & & & \\ \cdot & & & & \\ \cdot & & & & \\ a_{n1} & a_{n2} & \cdots & a_{nn} & a_{n,n+1} \end{bmatrix},$$

Construct matrix B:

$$B = \begin{bmatrix} b_{11} & b_{12} & \cdots & b_{1n} & b_{1,n+1} \\ b_{21} & b_{22} & \cdots & b_{2n} & b_{2,n+1} \\ \cdot & & & & \\ \cdot & & & & \\ \cdot & & & & \\ b_{n1} & b_{n2} & \cdots & b_{nn} & b_{n,n+1} \end{bmatrix}$$

in the following manner:

First column:	`DO I=1,N`	`B(I,1)=A(I,1)`
First row:	`DO J=2,N+1`	`B(1,J)=A(1,J)/B(1,1)`
Second column:	`DO I=2,N`	`B(I,2)=A(I,2)-B(I,1)*(1,2)`

Second row: `DO J=3,N+1   B(2,J)=(A(2,J)-B(2,1)*B(1,J))/B(2,2)`

Third column: `DO I=3,N   B(I,3)=A(I,3)-B(I,1)*B(1,3)-B(I,2)*B(2,3)`

Third row: `DO J=4,N+1   B(3,J)=(A(3,J)-B(3,1)*B(1,J))-B(3,2)*(2,J)/B(3,3)`

and so forth.

The coefficients of the equivalent triangular system have 0 below the diagonal, 1 on the diagonal, and b_{ij}, $i = 1, 2, 3, \ldots, n$, $j > i$, above the diagonal. For $n = 3$, the equivalent triangular system is

$$\begin{aligned} x_1 + b_{12}x_2 + b_{13}x_3 &= b_{14} \\ x_2 + b_{23}x_3 &= b_{24} \\ x_3 &= b_{34} \end{aligned}$$

From matrix B, the value of the determinant of the system can be found:

$$D = \prod_{i=1}^{n} b_{ii}$$

8.3.1 Find the exact solution of the linear system defined by matrix A. Evaluate the determinant of the left-hand coefficients.

$$A = \begin{bmatrix} 3 & -1 & -1 & -8 \\ 1 & 3 & -1 & 2 \\ 1 & 1 & 3 & 10 \end{bmatrix}$$

$$B = \begin{bmatrix} 3 & -\frac{1}{3} & -\frac{1}{3} & -\frac{8}{3} \\ 1 & \frac{10}{3} & \frac{1}{5} & \frac{7}{5} \\ 1 & \frac{4}{3} & \frac{18}{5} & 3 \end{bmatrix} \Rightarrow \begin{aligned} x_1 - \tfrac{1}{3}x_2 - \tfrac{1}{3}x_3 &= -\tfrac{8}{3} \\ x_2 - \tfrac{1}{5}x_3 &= \tfrac{7}{5} \\ x_3 &= 3 \end{aligned}$$

$$x = (-1, 2, 3)$$

and

$$D = \begin{vmatrix} 3 & -1 & -1 \\ 1 & 3 & -1 \\ 1 & 1 & 3 \end{vmatrix} = (3)(\tfrac{10}{3})(\tfrac{18}{5}) = 36$$

8.3.2 Write a program to solve a 4×4 linear system by the Gauss–Crout method. Also print out the value of the determinant D.

Test your program on

$$A = \begin{bmatrix} .20 & .32 & .12 & .30 & .94 \\ .10 & .15 & .24 & .32 & .81 \\ .20 & .24 & .46 & .36 & 1.26 \\ .60 & .40 & .32 & .24 & 1.52 \end{bmatrix}$$

Check your program by also printing out the "residues":

$$R_i = \sum_{j=1}^{4} a_{ij}x_j - a_{i5}, \qquad i = 1, 2, 3, 4$$

8.3.3 For the circuit network shown in Figure 8.6 with indicated currents i_k and resistances R_k, where $k = 1, 2, 3, 4, 5$:

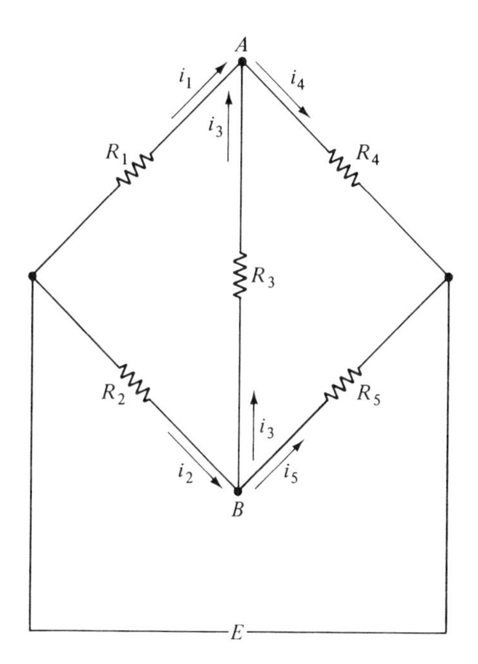

Fig. 8.6

Kirchhoff's laws state:

At point A	$i_1 + i_3 = i_4$
At point B	$i_2 = i_3 + i_5$
Around the left loop	$R_1 i_1 + R_2 i_2 + R_3 i_3 = 0$
Around the right loop	$R_4 i_4 + R_5 i_5 + R_3 i_3 = 0$
Around the outside loop	$R_1 i_1 + R_4 i_4 = E$

Substitute the first two equations in the last three to reduce this to three linear equations. Then write a program to read in R_1, R_2, R_3, R_4, R_5, and E and print out values of i_1, i_2, i_3, i_4, i_5.

8.4 The Gauss–Seidel Solution

A modification of the example of the iterative scheme indicated at the beginning of this chapter is the Gauss–Seidel method. For the same example and initial estimate $x_1^{(0)} = x_2^{(0)} = x_3^{(0)} = 0$, the iteration is determined by

$$\begin{aligned} x_1^{(k+1)} &= \tfrac{1}{3}[-8 + x_2^{(k)} + x_3^{(k)}] \\ x_2^{(k+1)} &= \tfrac{1}{3}[\;2 - x_1^{(k+1)} + x_3^{(k)}], \qquad k = 0, 1, 2, \ldots \\ x_3^{(k+1)} &= \tfrac{1}{3}[\;10 - x_1^{(k+1)} - x_2^{(k+1)}] \end{aligned}$$

For the iterative solution in Section 8.1, an initial estimate of the solution $[x_1^{(0)}, x_2^{(0)}, x_3^{(0)}]$ was used to compute the entire next approximation of the solution $[x_1^{(1)}, x_2^{(1)}, x_3^{(1)}]$. In the Gauss–Seidel algorithm an initial estimate is used to compute the first element of the next approximation $x_1^{(1)}$, and this with the remainder of the initial estimate is used to compute $x_2^{(1)}$, and so forth.

A few iterations found by slide rule are as follows:

k	x_1	x_2	x_3
0	0	0	0
1	−2.667	+1.556	+3.704
2	−.914	+2.206	+2.903
3	−.964	+1.956	+3.003
	↓	↓	↓
	−1	2	3

Again, it is at least plausible that this sequence converges to the solution and does so more rapidly than the algorithm suggested in Section 8.1.

A sufficient condition for convergence of the Gauss–Seidel iteration is that the system have a *dominant diagonal.* That is, if we can rearrange rows so that the absolute value of every diagonal element is greater than the sum of absolute values of all other elements on the same row, of left-hand coefficients, then the Gauss–Seidel iteration will converge. In general, if the linear system

$$\sum_{j=1}^{n} a_{ij}x_j = c_i, \qquad i = 1, 2, \ldots, n$$

is such that

$$|a_{ii}| > \sum_{\substack{j=1 \\ j\neq i}}^{n} |a_{ij}|, \qquad i = 1, 2, \ldots, n$$

then the sequence $[x_1^{(k)}, \ldots, x_n^{(k)}]$, where $[x^{(0)}, \ldots, x_n^{(0)}]$, is any given estimate and

$$x_i^{(k+1)} = \frac{1}{a_{ii}}\left[b_i - \sum_{\substack{j=1 \\ j\neq i}}^{n} a_{ij}x_j^{(k+p)}\right], \qquad i = 1, 2, \ldots, n;$$

$$\text{where } k = 1, 2, \ldots, \qquad \text{and } p = \begin{cases} 0 & j > i \\ 1 & j < i \end{cases},$$

converges to the solution.

The attractive simplicity of such a solution method is clouded by its restriction to systems for which this method converges. Even with this condition met, the computed sequence may converge with unacceptable slowness. However, there are important classes of problems that meet the condition of convergence and for which this method is preferable to direct methods.

8.4.1 Write a program to solve a 4×4 linear system by the Gauss–Seidel iteration. Print out each iterate. As a test for stopping use

$$\sum_{i=1}^{4} [x_i^{(k)} - x_i^{(k-1)}]^2 < E^2 \quad \text{or} \quad k > 20$$

where E is a read-in error bound and k is a bound on the number of iterations. Test your program on:

$$\begin{aligned} 4x_1 - x_2 \qquad - x_4 &= 3 \\ -x_1 + 4x_2 - x_3 \qquad &= 3, \qquad E = .001 \\ - x_2 + 4x_3 - x_4 &= 7, \qquad x_1^{(0)} = x_2^{(0)} = x_3^{(0)} = x_4^{(0)} = 0 \\ -x_1 \qquad - x_3 + 4x_4 &= 9 \end{aligned}$$

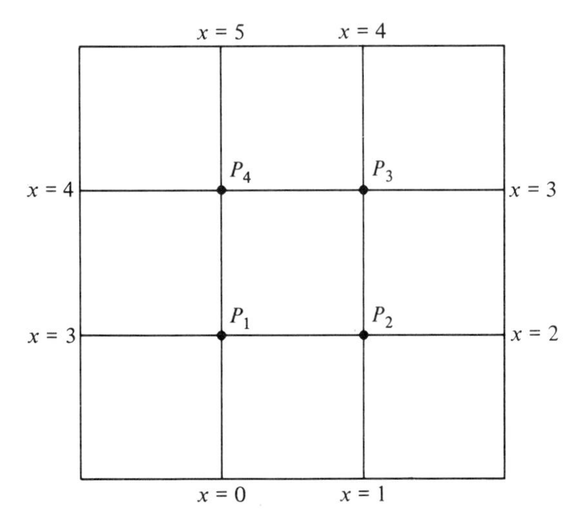

Fig. 8.7

The solution of this system of equations approximates the temperatures x_1, x_2, x_3, x_4 at the corresponding points P_1, P_2, P_3, P_4 on Figure 8.7, representing a heat conducting plate having the indicated temperatures maintained at its boundaries.

9

POLYNOMIAL APPROXIMATION

A given function $f(x)$ is said to be approximated over an interval by polynomial $P_n(x)$ with error $R(x)$ if

$$f(x) = P_n(x) + R(x)$$

The approximating polynomial $P_n(x)$ may be defined in many ways, and for many purposes, by appropriate conditions placed on the error $R(x)$.

9.1 Taylor Approximation

A polynomial of degree n approximating a given function $f(x)$ in the neighborhood of $x = a$ is defined by requiring that

$$R^{(k)}(a) = 0, \qquad k = 0, 1, 2, \ldots, n$$

Assuming that the first n derivatives of $f(x)$ exist,

$$f^{(k)}(x) = P_n^{(k)}(x) + R^{(k)}(x), \qquad k = 0, 1, \ldots, n$$

If $R^{(k)}(a) = 0$, where $k = 0, 1, \ldots, n$, then

$$f^{(k)}(a) = P_n^{(k)}(a), \qquad k = 0, 1, \ldots, n$$

Thus the approximating polynomial $P_n(x)$ and $f(x)$ have the same value at $x = a$, and the first n derivatives of the polynomial are equal to the corresponding derivatives of $f(x)$ at $x = a$.

How can this approximating polynomial be constructed? Let

$$P_n(x) = A_0 + A_1(x - a) + A_2(x - a)^2 + \cdots + A_n(x - a)^n$$

$$P_n(a) = f(a) = A_0$$

$$P_n'(x) = A_1 + 2!A_2(x - a) + 3A_3(x - a)^2 + \cdots + nA_n(x - a)^{n-1}$$

$$P_n'(a) = f'(a) = A_1$$

$$P_n''(x) = 2!A_2 + 3.2A_3(x - a) + \cdots + n(n - 1)A_n(x - a)^{n-2}$$

$$P_n''(a) = f''(a) = 2!A_2$$

$$\vdots$$

$$P_n^{(n)}(x) = \qquad\qquad + n!A_n$$

$$P_n^{(n)}(a) = f^{(n)}(a) = n!A_n$$

Thus

$$P_n(x) = f(a) + f'(a)(x - a) + \frac{f''(a)}{2!}(x - a)^2 + \cdots + \frac{f^{(n)}(a)}{n!}(x - a)^n$$

Taylor's theorem shows that the error, or remainder, can be expressed in the form

$$R(x) = \frac{f^{(n+1)}(\zeta)}{(n + 1)!}(x - a)^{n+1}, \qquad |\zeta - a| < |x - a|$$

In the following problems we will be concerned merely with the construction of the polynomial and not with its merit as an approximation of $f(x)$. [A study of the error $R(x)$, with numerous examples, can be found in most calculus text books.]

9.1.1 Construct a polynomial of degree 3 approximating $\tan^{-1} x$ at $x = 0$ (Fig. 9.1) in the Taylor sense:

$$f(x) = \tan^{-1} x \qquad f(0) = 0$$

$$f'(x) = \frac{1}{1 + x^2} \qquad f'(0) = 1$$

$$f''(x) = \frac{2x}{(1 + x^2)^2} \qquad f''(0) = 0$$

$$f'''(x) = \frac{2(3x^2 - 1)}{(1 + x^2)^3} \qquad f'''(0) = -2$$

9.1.2 Estimate the value of $\int_0^{.25} \sqrt{x} \tan^{-1} x \, dx$ using the approximating polynomial $P_3(x)$ found in 9.1.1.

$$\int_0^{.25} \sqrt{x} \tan^{-1} x \, dx \simeq \int_0^{.25} (x^{3/2} - \tfrac{1}{3}x^{7/2}) \, dx$$
$$= (\tfrac{2}{5}x^{5/2} - \tfrac{2}{27}x^{9/2}) \Big|_0^{.25} \simeq .012$$

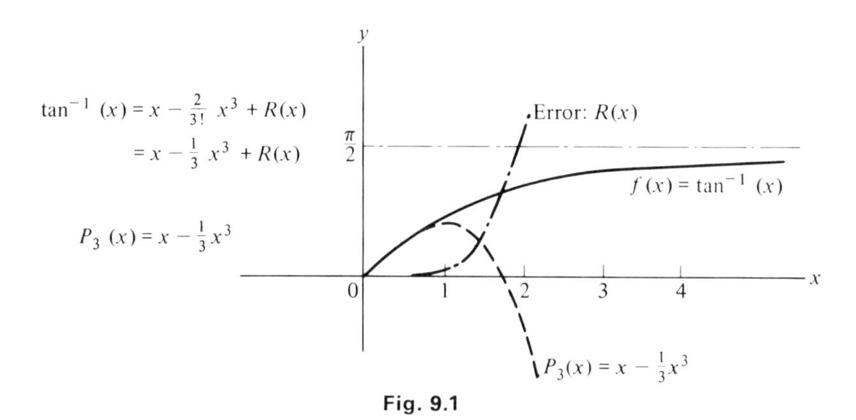

Fig. 9.1

9.1.3 Consider the functions and corresponding approximating polynomials given in Section 4.11. In each case find the approximating polynomial of next higher degree.

9.1.4 For the polynomial function of the problem in 4.8.1,

$$f(x) = x^3 + 2x^2 + 10x - 20$$

find an approximating polynomial of degree 2 in the neighborhood of $x = 1.3$. Find the root of $P_2(x)$ near 1.3 by the quadratic formula. Compare this result to that of the problem in 4.8.4.

9.1.5 Estimate the value of $\int_0^{.1} e^{-x} \, dx$ by replacing the integrand by an approximating polynomial in the Taylor sense at $x = 0$, of degree 3. Compare this to the tabulated value of $1 - e^{-.1} = +.09516258$.

In a similar manner estimate $\int_0^{.1} e^{-x^2} \, dx$. Can you use again the *funda-*

mental theorem of calculus to compare this estimate with tabulated values of elementary functions?

9.1.6 The equation

$$\frac{x}{2} - e^{-|x-1|} = 0$$

has a root near $x = 1.3$. Replace $e^{-|x-1|}$ by a quadratic polynomial, approximating it at $x = 1.3$. Solve, by quadratic formula, the resulting equation so as estimate the root of this equation.

9.1.7 Find the quadratic polynomial approximating $\log(1 + x)$ in the Taylor sense at $x = 0$. For $x_k = -.1 + k(.01)$, where $k = 0, 1, \ldots, 20$, print out the Fortran evaluation of $\log(1 + x)$, $P_2(x)$, and the error $R(x) = \log(1 + x) - P_2(x)$.

9.2 Polynomial Interpolation

A given function is said to be interpolated over an interval containing $x_0, x_1, \ldots, x_n$ by polynomial $P_n(x)$ with error $R(x)$ if

$$f(x) = P_n(x) + R(x)$$

and $R(x_k) = 0$, where $k = 0, 1, 2, \ldots, n$.
If we are given for each of the $n + 1$ distinct values $x_0, x_1, \ldots, x_n$ the corresponding functional values $f_0, f_1, \ldots, f_n$, the interpolating polynomial is such that

$$f_k = P_n(x_k), \qquad k = 0, 1, \ldots, n$$

An illustration for $n = 3$ is shown in Figure 9.2.

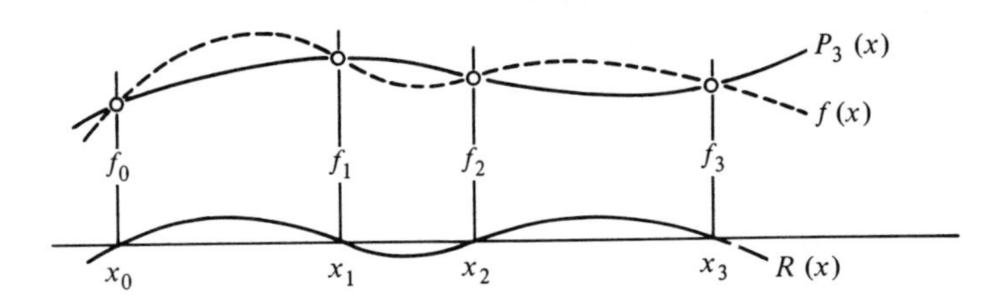

Fig. 9.2

A text on numerical analysis would show that the error can be expressed as

$$R(x) = \frac{f^{(n+1)}(\zeta)}{(n+1)!}(x - x_0)(x - x_1) \cdots (x - x_n), \qquad x_0 < \zeta < x_n$$

Again, we will be concerned in the following problems with the construction of the interpolating polynomial and not with quantitative estimates of error.

This problem may be restated: Find the polynomial of degree (at most) n whose graph contains $n + 1$ given points. There are several ways of constructing

$$P_n(x) = a_n x^n + a_{n-1}x^{n-1} + \cdots + a_1 x + a_0$$

such that $P_n(x_k) = f_k$, where $k = 0, 1, \ldots, n$. We may find the unknown coefficients $a_n, a_{n-1}, \ldots, a_0$ by solving the linear system

$$\sum_{k=0}^{n} x_i^k a_k = f_i, \; i = 0, 1, \ldots, n \tag{1}$$

For $n = 2$ this system is

$$\begin{aligned}
(x_0^2)a_2 + (x_0)a_1 + a_0 &= f_0 \\
(x_1^2)a_2 + (x_1)a_1 + a_0 &= f_1 \\
(x_2^2)a_2 + (x_2)a_1 + a_0 &= f_2
\end{aligned}$$

We will show in Section 9.3 that if $x_0, x_1, \ldots, x_n$ are distinct, the nonhomogeneous system (1) is consistent; that is, a solution $(a_n, a_{n-1}, \ldots, a_n)$ exists which is also unique.

It will then follow that the Vandermonde determinant does not vanish if $x_0, x_1, \ldots, x_n$ are distinct:

$$V = \begin{vmatrix} x_0^n & x_0^{n-1} & \cdots & x_0 & 1 \\ x_1^n & x_1^{n-1} & \cdots & x_1 & 1 \\ \cdot & & & & \\ \cdot & & & & \\ \cdot & & & & \\ x_n^n & x_n^{n-1} & \cdots & x_n & 1 \end{vmatrix} \neq 0$$

9.2.1 Write a program to read in four points having distinct abscissas. Solve the corresponding linear system by the Gauss–Crout method developed in Section 8.3. Write out the coefficients of the interpolating polynomial.

9.2.2 For the four points of problem 9.2.1, use Fortran values of $\sin x_j$, $x_j = \pi/16 + (j)\pi/8$, where $j = 0, 1, 2, 3$. Print out values of $\sin x_k$ (Fortran evaluation), interpolated values $P_3(x_k)$, and the error $R(x_k) = \sin x_k - P_3(x_k)$ for $x_k = (k)\pi/64$, where $k = 0, 1, 2, \ldots, 32$.

9.2.3 For the four points of problem 9.2.1, use Fortran values of $\log(1 + x)$, where $x = -.5, .0, +.5, +1.5$. Print out values of $\log(1 + x_k)$ (Fortran evaluation), interpolated values $P_3(x_k)$, and the error $R(x_k) = \log(1 + x_k) - P_3(x_k)$ for $x_k = -.5 + (k)(.1)$, where $k = 0, 1, 2, \ldots, 20$.

9.3 The Lagrange Form

We can show the existence of the interpolating polynomial of degree n (at most), given $P_n(x_k) = f_k$, where $k = 0, 1, \ldots, n$, and at the same time present an interesting and useful construction of this polynomial. For an illustration we will consider the case $n = 2$.

Let

$$P_2(x) = L_0(x)f_0 + L_1(x)f_1 + L_2(x)f_2$$

where

$$L_0(x) = \frac{(x - x_1)(x - x_2)}{(x_0 - x_1)(x_0 - x_2)}$$

$$L_1(x) = \frac{(x - x_0)(x - x_2)}{(x_1 - x_0)(x_1 - x_2)}$$

$$L_2(x) = \frac{(x - x_0)(x - x_1)}{(x_2 - x_0)(x_2 - x_1)}$$

The Lagrange coefficient polynomials have the property

$$L_k(x_j) = \begin{cases} 0, & j \neq k \\ 1, & j = k \end{cases}$$

and their graphs (for $n = 2$) are shown in Figure 9.3.

Note that $P_2(x) = \sum_{k=0}^{2} L_k(x)f_k$ has the following properties:

1. It is the sum of quadratic polynomials and is thus a quadratic polynomial.

2. $P_2(x_k) = f_k$, where $k = 0, 1, 2$.

By this construction we have shown the existence of a polynomial of degree 2 containing three given points. A polynomial of degree n containing

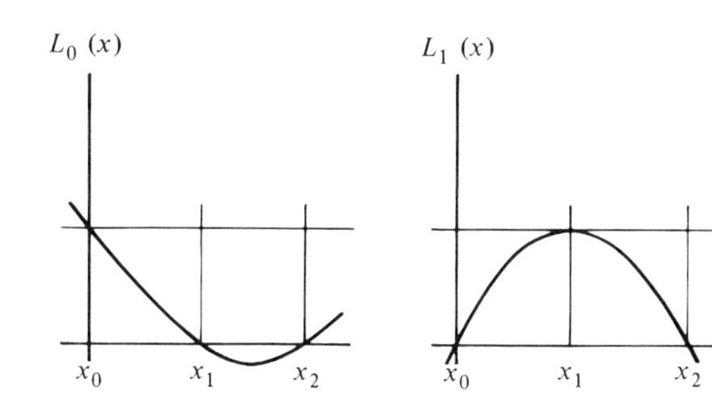

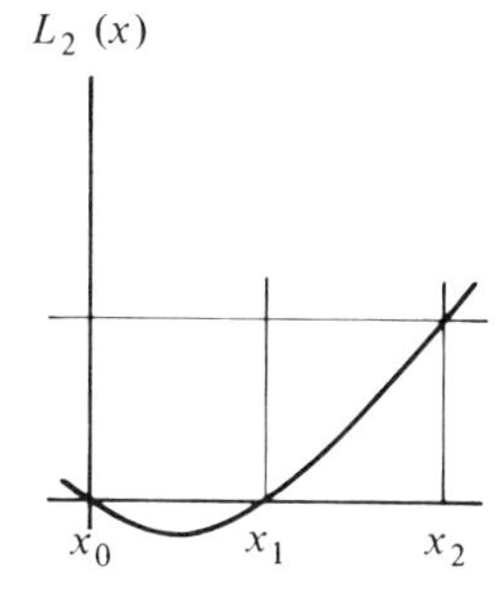

Fig. 9.3

the $n + 1$ given points (x_k, f_k), where $k = 0, 1, \ldots, n$ is

$$P_n(x) = \sum_{k=0}^{n} L_k(x) f_k$$

where

$$L_k(x) = \frac{(x - x_0)(x - x_1) \cdots (x - x_{k-1})(x - x_{k+1}) \cdots (x - x_n)}{(x_k - x_0)(x_k - x_1) \cdots (x_k - x_{k-1})(x_k - x_{k+1}) \cdots (x_k - x_n)}$$

Since each $L_k(x)$ is a polynomial of degree n, and since

$$L_k(x_j) = \begin{cases} 0, & j = 0, 1, \ldots, \text{n}, \quad j \neq k \\ 1, & j = k \end{cases}$$

$P_n(x)$ is a polynomial of degree n such that

$$P_n(x_k) = f_k, \qquad k = 0, 1, \ldots, n$$

9.3.1 Determine the cubic polynomial containing $(0, -1)$, $(1, 2)$, $(2, 1)$, $(3, 0)$.

$$\begin{aligned} P_3(x) &= \frac{(x-1)(x-2)(x-3)}{(-1)(-2)(-3)}(-1) + \frac{x(x-2)(x-3)}{(1)(-1)(-2)}(2) \\ &\quad + \frac{x(x-1)(x-3)}{(2)(1)(-1)} \qquad (1) \\ &= \tfrac{2}{3}x^3 - 4x^2 + \tfrac{19}{3}x - 1 \end{aligned}$$

These coefficients could also be obtained by the program developed in problem 9.2.1 for the solution of the four linear conditions on unknown polynomial coefficients.

9.3.2 For an interpolating polynomial of degree 3, use a Fortran statement subroutine to define the Lagrange coefficient polynomial:

```
L(X,A,B,C,D)=(X-B)*(X-C)*(X-D)/((A-B)*(A-C)*(A-D))
```

Program the evaluation of the four-point interpolating polynomial for $(x_k, \sin x_k)$, $x_k = (k)(\pi/6)$, where $k = 0, 1, 2, 3$. Print out Fortran values of $\sin x_j$, the values of $P_3(x_j)$, and the error $R(x_j) = \sin x_j - P_3(x_j)$ for $x_j = (j)(\pi/60)$, where $j = 0, 1, 2, \ldots, 30$.

9.3.3 Draw the graphs of each of the polynomials

$$L_{-1}(s) = -\frac{s(s-1)(s-2)}{3!}$$

$$L_0(s) = +\frac{(s+1)(s-1)(s-2)}{(2!)(1!)}$$

$$L_1(s) = -\frac{(s+1)(s)(s-2)}{(2!)(1!)}$$

$$L_2(s) = +\frac{(s+1)(s)(s-1)}{3!}$$

Using statement subroutines to define $L_k(s)$, where $k = -1, 0, 1, 2$, construct the cubic interpolating polynomial for $f(s) = e^{-s^2}$ having zero error at $s = -1, 0, 1, 2$:

$$P_3(s) = L_{-1}(s)f_{-1} + L_0(s)f_0 + L_1(s)f_1 + L_2(s)f_2$$

Use the Fortran subroutine for values of f_k = EXP(−S*S), where $s = k$. Print out, for $s = 0, .1, .2, \ldots, 1.0$ values of EXP(−S*S), $P_3(s)$, and the error $R(s)$.

9.3.4 The cross section of a road surface is to be parabolic and, at a curve, banked (Fig. 9.4).

Construct a program to read in width (W feet), bank (B feet), crown (C feet), and an odd number of stakes (N). Stakes are to be placed with equal horizontal spacing across the road to identify the parabolic surface. Write

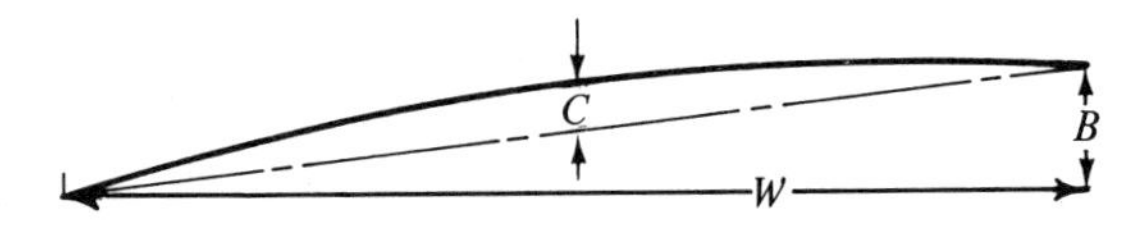

Fig. 9.4

out the stake number, its distance from the inside edge of the road, and the height of the stake above the level line AA. As an example, let $W = 40$, $B = 1$, $C = \frac{1}{3}$, and $N = 11$.

9.3.5 Find the quadratic polynomial interpolating $\log(1 + x)$ with zero error at $x = -.5, 0, +.5$, and print out for $x_k = -.1 + (k)(.01)$, where $k = 0, 1, \ldots, 20$, the Fortran value of $\log(1 + x)$, the interpolated values of $P_2(x)$, and the error $R(x) = \log(1 + x) - P_2(x)$.

Repeat this if the polynomial has zero error at $x = -.2, .0, +.2$ and again with zero error at $-.1, .0, +.1$. Compare these tabulations of error with those of problem 9.1.5, where the approximating polynomial was defined by

$$R(0) = R'(0) = R''(0) = 0$$

9.4 Polynomial Fits

Consider the problem of determining a polynomial of degree 2 whose graph passes through (a) Less than three given points; (b) Exactly three given points; and (c) More than three given points.

(a) There are infinite sets of polynomials of degree 2 whose graph contains less than three given points (Fig. 9.5). In each case the system of equations on the coefficients of $P_2(x) = a_2x^2 + a_1x + a_0$ are "underdetermined":

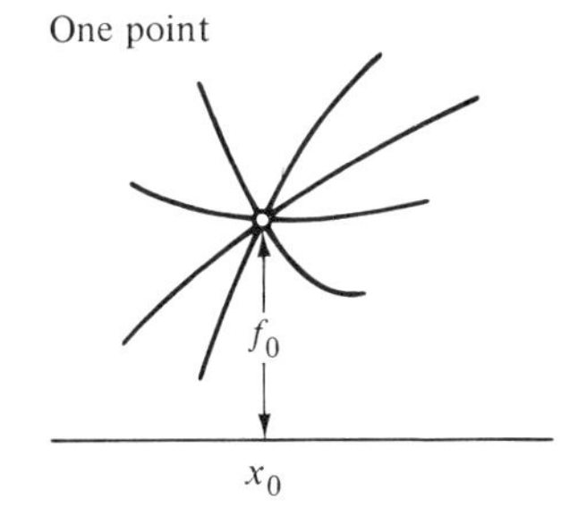

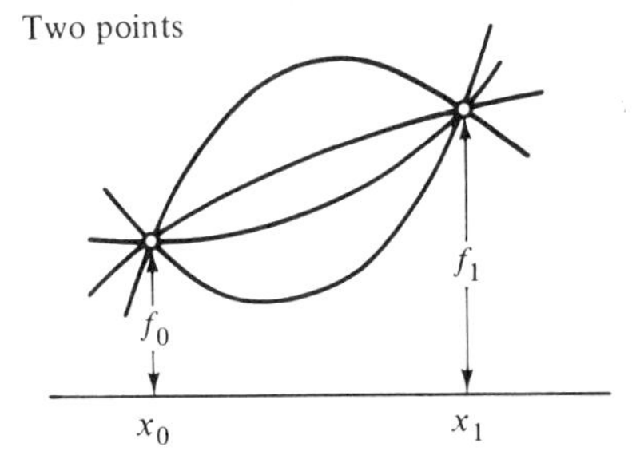

Fig. 9.5

$$a_2x_0^2 + a_1x_0 + a_0 = f_0, \qquad\qquad \begin{aligned} a_2x_0^2 + a_1x_0 + a_0 &= f_0 \\ a_2x_1^2 + a_1x_1 + a_0 &= f_1 \end{aligned}$$

The equivalent conditions expressed in terms of error are

$$R(x_0) = 0 \qquad\qquad R(x_k) = 0, \quad k = 0, 1$$

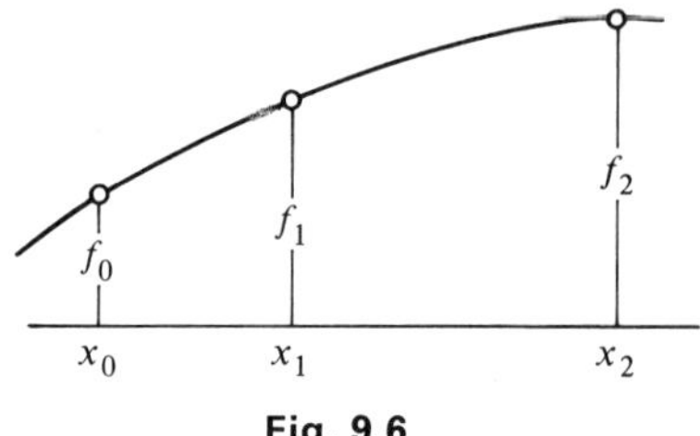

Fig. 9.6

(b) If x_0, x_1, x_2 are all distinct, there is one and only one polynomial of degree 2 whose graph (Fig. 9.6) contains the points (x_i, f_i), $i = 0, 1, 2$.

The conditions on the coefficients are

$$\sum_{j=0}^{2} a_j x_k^j = f_k, \qquad k = 0, 1, 2$$

or

$$R(x_k) = 0, \qquad k = 0, 1, 2$$

If the given points are on a line, $a_2 = 0$. If the given points are on a horizontal line, $a_2 = a_1 = 0$.

(c) There is, in general, no polynomial of degree 2 whose graph contains more than three given points (Fig. 9.7).

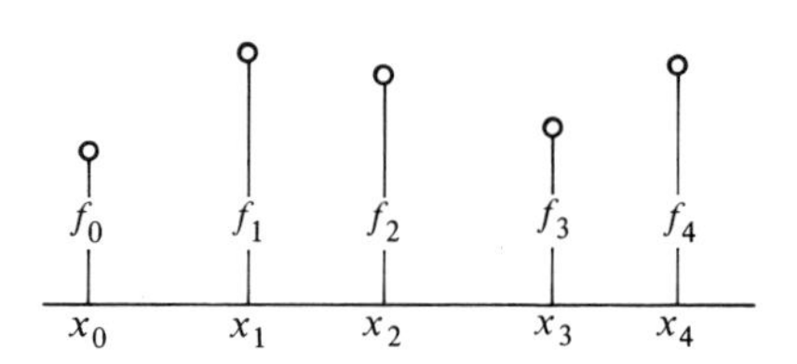

Fig. 9.7

The conditions on the coefficients

$$\sum_{j=0}^{2} a_j x_k^j = f_k, \qquad k = 0, 1, 2, \ldots, m, \quad m > 2$$

represent an "overdetermined" system.

We may extend these considerations to state that

1. There are infinite sets of polynomials of degree n, the graph of each containing m given points, where $m < n + 1$.

2. There is a unique polynomial of degree n that interpolates exactly at $n + 1$ given points.

3. There is, in general, no polynomial of degree n that interpolates exactly at all of m given points, $m > n + 1$.

Suppose that we have the results of many trials of an experiment in which a value of x is assigned and a corresponding value of some related variable, $f(x)$, is determined. Also suppose that there may be evidence to support the belief that $f(x)$ is actually a polynomial or may be usefully represented by a polynomial. If, for example, the polynomial function is of degree 2, should we throw away all the experimental evidence except three of the points to determine $P_2(x)$? A more reasonable approach is to "fit" a polynomial to all the points by some appropriate criterion.

Of the many possible criteria that will fit a polynomial to a given set of data, one that has useful properties, particularly in problems of statistical inference, is the *least-squares fit*.

Let the deviation of f_k from the corresponding $P_n(x_k)$ be $D(x_k)$, (Fig. 9.8), where

$$D(x_k) = f_k - P_n(x_k), \qquad k = 1, 2, 3, \ldots, m$$

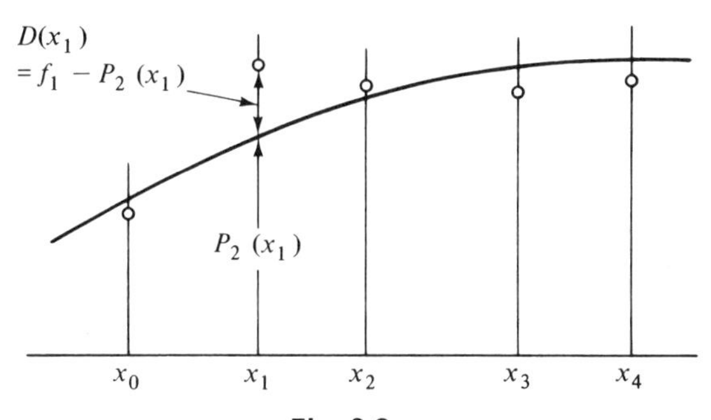

Fig. 9.8

The least-squares condition determining the coefficients of the fitted polynomial $P_n(x)$ is

$$\sum_{k=1}^{m} [D(x_k)]^2 \bigg|_{\min} = \sum_{k=1}^{m} [f_k - (a_n x_k^2 + \cdots + a_0)]^2 \bigg|_{\min}$$

$$= G(a_n, a_{n-1}, \ldots, a_0) \bigg|_{\min}$$

Let us assume that the function G has only one extreme value and that it is a minimum. For the example $n = 2$,

$$G(a_2, a_1, a_0) = \sum_{k=1}^{m} [f_k - (a_2 x_k^2 + a_1 x_k + a_0)]^2$$

G is minimum if

$$\frac{\partial G}{\partial a_2} = \frac{\partial G}{\partial a_1} = \frac{\partial G}{\partial a_0} = 0$$

The symbol $\partial G/\partial a_2$ is read "the partial derivative of G with respect to a_2" and is the derivative of G treating a_2 as the sole independent variable.

$$\frac{\partial G}{\partial a_0} = 2 \sum_{k=1}^{m} [f_k - a_2 x_k^2 - a_1 x_k - a_0](-1) = 0$$

$$\frac{\partial G}{\partial a_1} = 2 \sum_{k=1}^{m} [f_k - a_2 x_k^2 - a_1 x_k - a_0](-x_k) = 0$$

$$\frac{\partial G}{\partial a_2} = 2 \sum_{k=1}^{m} [f_k - a_2 x_k^2 - a_1 x_k - a_0](-x_k^2) = 0$$

These three linear conditions on the unknown coefficients simplify to the form

$$S_{x^2} a_2 + S_x a_1 + m a_0 = S_y$$
$$S_{x^3} a_2 + S_{x^2} a_1 + S_x a_0 = S_{xy}$$
$$S_{x^4} a_2 + S_{x^3} a_1 + S_{x^2} a_0 = S_{x^2 y}$$

where

$$S_x = \sum_{k=1}^{m} x_k, \quad S_{x^2} = \sum_{k=1}^{m} x_k^2, \quad \ldots, \quad S_{x^2 y} = \sum_{k=1}^{m} x_k^2 y_k$$

9.4.1 Find the least-squares fit of a polynomial of degree 2 to the given points $(-1, 0)$, $(0, 1)$, $(1, 0)$, $(2, 0)$ (Fig. 9.9).

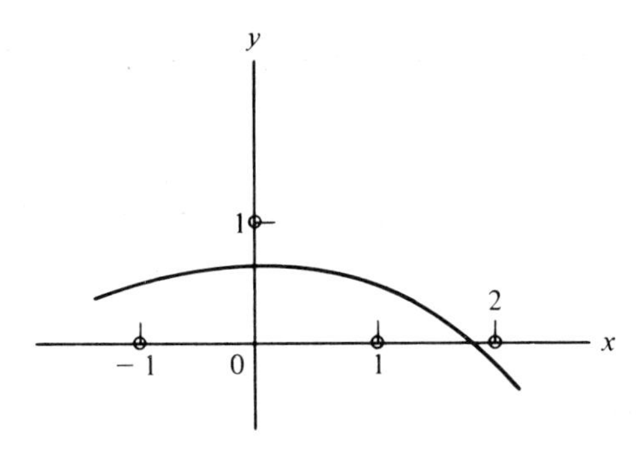

Fig. 9.9

	x	y	x^2	xy	x^3	x^2y	x^4
	−1	0	1	0	−1	0	1
	0	1	0	0	0	0	0
	1	0	1	0	1	0	1
	2	0	4	0	8	0	16
Σ	2	1	6	0	8	0	18

$$6a_2 + 2a_1 + 4a_0 = 1$$
$$8a_2 + 6a_1 + 2a_0 = 0$$
$$18a_2 + 8a_1 + 6a_0 = 0$$
$$a_2 = -\tfrac{5}{20}, \quad a_1 = \tfrac{3}{20}, \quad a_0 = \tfrac{11}{20}$$

or

$$P_2(x) = \tfrac{1}{20}(-5x^2 + 3x + 11)$$

If $n = 1$ and $m > 2$, this linear polynomial of least-squares fit is called *the line of regression* in statistical analysis. The conditions on the coefficients of $P_1(x) = a_1x + a_0$ are

$$S_x a_1 + m a_0 = S_y$$
$$S_{x^2} a_1 + S_x a_0 = S_{xv}$$

9.4.2 Derive the linear conditions imposed on the coefficients of $P_1(x)$ by its least-squares fit to $m > 2$ points. Show that the point $(\bar{x}, \bar{y})$, $\bar{x} = (1/n)S_x$, $\bar{y} = (1/n)S_y$ lies on the graph of $P_1(x)$.

9.4.3 Find the line of regression for the data (0, 0), (1, 1), (2, 2), (3, 1) by hand computation.

9.4.4 Write a program to read in $N > 2$ number pairs, using a trailer card to indicate the last of the data. Tabulate for each x_k the corresponding f_k and value of the fitted $P_1(x_k)$. Write out the equation of the line of regression and the sum of the squares of the data deviations from $P_1(x)$.

9.4.5 Write a program to read in $m > 4$ data points and print out the coefficients of the polynomial of degree 3 having a least-squares fit. Use the program of problem 8.4.1 in the form of a subroutine to solve the linear system.

10

NUMERICAL INTEGRATION

A major thrust of a basic calculus course is the definition of the definite integral and its evaluation through the fundamental theorem of calculus. For instance, we find the following notation and algorithm both familiar and perhaps of mystifying simplicity:

$$\int_0^1 x^3\,dx = \frac{x^4}{4}\bigg|_0^1 = \frac{(1)^4}{4} - \frac{(0)^4}{4} = \frac{1}{4}$$

Why did we select the function $x^4/4$ for this role, and why do we claim that the value of this limit is the difference of the values of $x^4/4$ at $x = 1$ and $x = 0$? The answers to these questions are clearly set forth in most calculus textbooks.

We might also desire to find the limit

$$\int_0^1 e^{-x^2}\,dx$$

Following the directive of the fundamental theorem of calculus we consult our catalog of derivatives, searching for a function whose derivative is e^{-x^2}, but meet with no success. What then? If we know the properties of the functions $u = x^2$ and e^{-u} and render a sketch of e^{-x^2} over [0, 1] (Fig. 10.1), it is

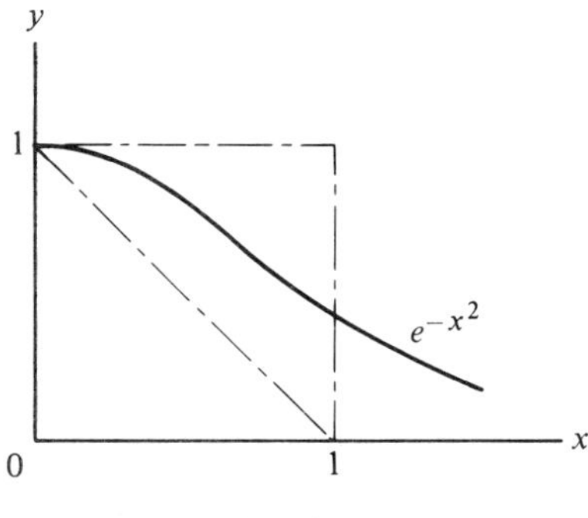

Fig. 10.1

not only obvious that this limit exists but that there are evident bounds on its value.

$$\tfrac{1}{2} < \int_0^1 e^{-x^2}\,dx < 1$$

It is right and proper to pay reverent attention to "formal integration." However, it is wrong to be left with the impression that, to evaluate an integral, it is necessary to produce the antiderivative of the integrand in a closed form involving elementary functions. In the application of mathematics it is likely that, for the integrals that we meet, antiderivatives will not exist in familiar formula form.

10.1 Sundry Ways and Means

In the interests of mind expansion, there are listed below a few of the multitude of procedures that approximate the number

$$\int_a^b f(x)\,dx$$

given a, b, and $f(x)$, integrable over $a \leqq x \leqq b$. There will be some numerical algorithms and exercises in Section 10.2. to approximate the definite integral.

The first five on this list involve a scale drawing of the figure bounded by $y = f(x)$, $y = 0$, $x = a$, $x = b$ (see region A in Figure 10.2).

10.1.1 The Paper Caper. On the same scale as the drawing, and on the same grade of paper construct a square of unit area (Region S on Figure 10.2). Cut out figure A and figure S and weigh both.

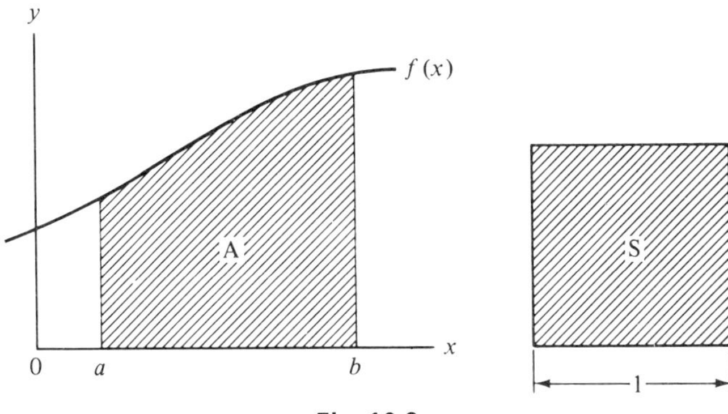

Fig. 10.2

$$\int_a^b f(x)\,dx \simeq \frac{\text{wt. of figure A}}{\text{wt. of figure S}}$$

10.1.2 Gallup's Gambit. Show the cutouts used in the above procedure to N people, N as large as feasible, and ask each to estimate the area of figure A. You might explain that area means the number of squares needed to just cover figure A and that, at least mentally, scissors may be used.

$$\int_a^b f(x)\,dx \simeq \frac{\text{sum of the estimates}}{N}$$

10.1.3 The Dartful Dodge. Paste figure A on a rectangular board of known area (Fig. 10.3) Consider the board subdivided into a grid of equal area regions. Throw darts at the board in such a way that each subdivision of the board is *equally likely* to be hit, and record the number of hits (h) and misses (m) of figure A.

$$\int_a^b f(x)\,dx \simeq \frac{h}{h+m}\ \text{(area of the board)}$$

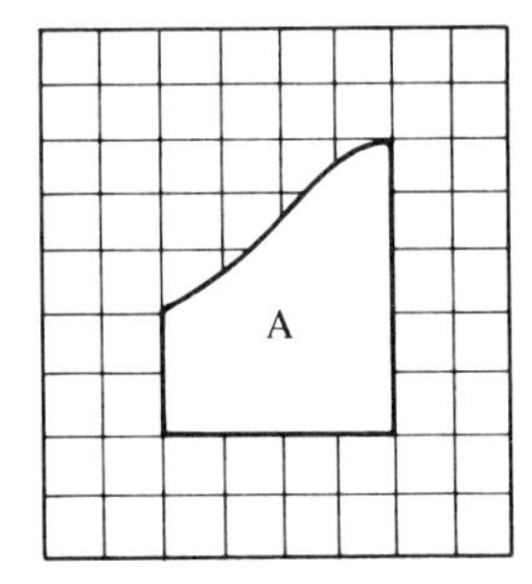

Fig. 10.3

There is a simpler way to play this game using a digital computer to simulate the "equally likely" method of dart throwing. Let C be an upper bound of $f(x)$ in $[a, b]$. A random-number-generator subroutine (refer Fig. 10.4 and problem 10.2.33) determines

$$0 \leqq \theta_i \leqq 1, \quad 0 \leqq \Phi_i \leqq 1$$

and the simulated dart strikes the point (x_i, y_i)

$$x_i = a + (b - a)\theta_i, \quad y_i = C\Phi_i$$

If $y_i \leqq f(x_i)$, a hit is counted, and if $y_i > f(x_i)$, a miss is counted. After many throws

$$\int_a^b f(x)\,dx \simeq \frac{h}{h + m}(b - a)C$$

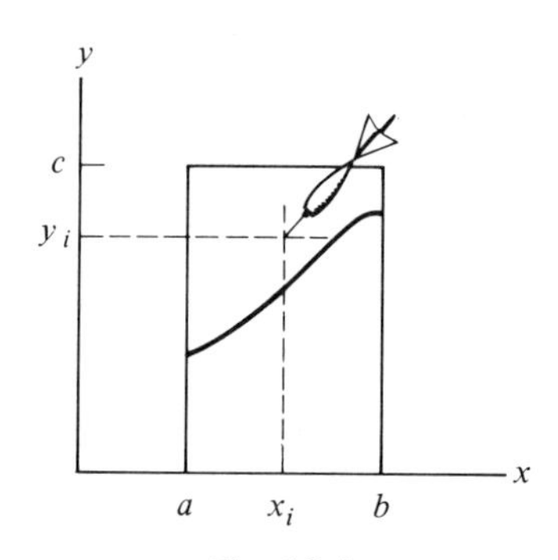

Fig. 10.4

It is intuitively evident that the error is "likely" to decrease as the number of throws increases. This procedure for approximating the value of a definite integral is appropriately called the *Monte Carlo method.*

10.1.4 The Planimeter. Ask a friendly civil engineer (over 30) to lend you his planimeter (Fig. 10.5) Establish the fixed point of the instrument and trace around the perimeter of figure A with the movable pointer. Read a number proportional to the area of figure A off the indicator dial of the planimeter. Repeat for figure S.

$$\int_a^b f(x)\,dx \simeq \frac{\text{reading for figure A}}{\text{reading for figure S}}$$

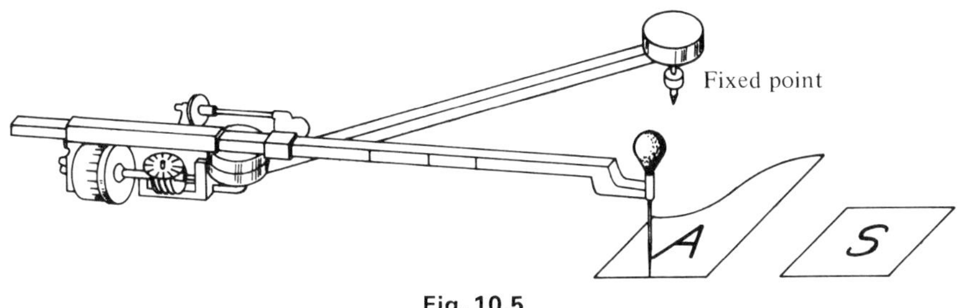

Fig. 10.5

The planimeter (there are various types) is a mechanical analog device to determine the area of a closed plane figure. The type shown in the illustration records on a dial a number proportional to the area swept out by the motion of a straight-line segment in a plane.

Consider the area swept out by segment QP of length l translated to $Q'P''$ and rotated to $Q'P'$ as shown in Figure 10.6:

$$\Delta \text{ area} = l\,\Delta n + \tfrac{1}{2}l^2\,\Delta\phi$$

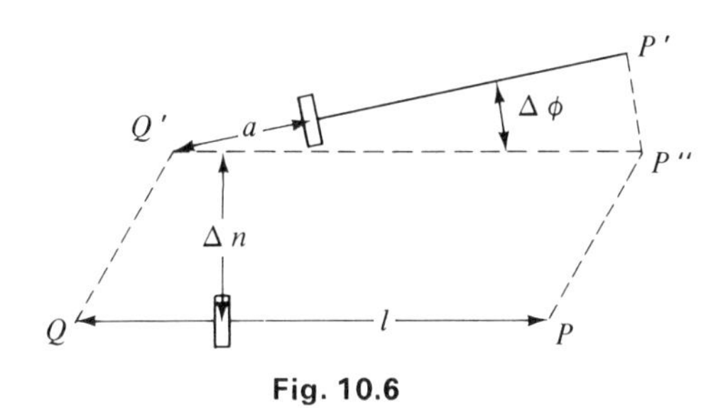

Fig. 10.6

A point on the circumference of the wheel rolls a distance Δs.

$$\Delta s = \Delta n + a\,\Delta\phi$$

$$\Delta \text{ area} = l\,\Delta s - al\,\Delta\phi + \tfrac{1}{2}l^2\,\Delta\phi$$

$$\text{area} = l\int_A ds - al\int_A d\phi + \tfrac{1}{2}l^2\int_A d\phi$$

If, as P traces the complete perimeter of figure A, the arm QP returns to its original position without making a complete revolution, $\int_A d\phi = 0$. Thus

$$\text{area} = l\cdot(\text{the net distance traversed by the roller})$$

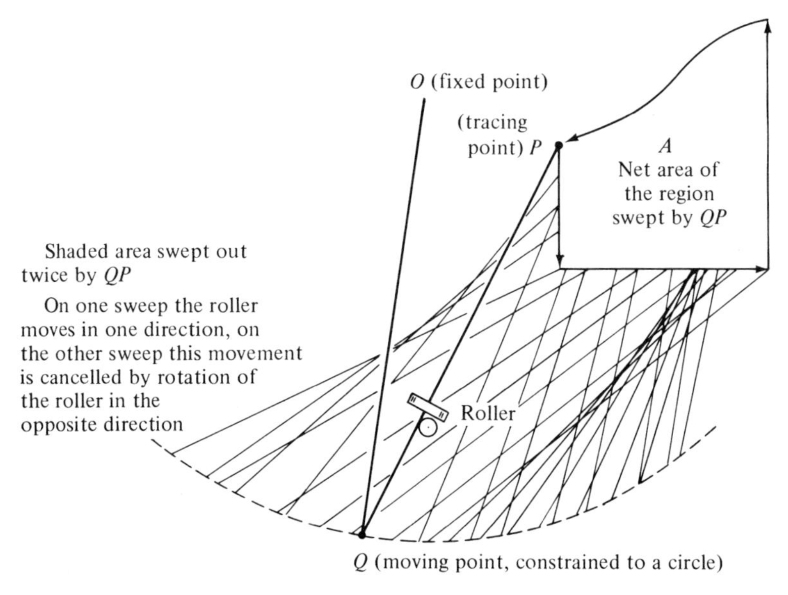

Fig. 10.7

In Figure 10.7, segment QP containing the roller is of length l. The planimeter has an arm OQ, with O a fixed point, constraining point Q to a circle.

10.1.5 A Drafting Dodge. Partition the interval $[a, b]$ of figure A into n subintervals. In each interval the draftsman selects a value of $f(x)$, say y_i, with visual effort to make the area of the rectangle $y_i(x_i - x_{i-1})$ equal to the area of the corresponding panel.

Of course,

$$\int_a^b f(x)\,dx = \sum_{i=1}^{n} y_i(x_i - x_{i-1}) + \text{error}$$

where "error" depends on the draftsman's visual skills. The following construction not only estimates the value of the definite integral but also gives an approximate construction of the graph of $F(x)$, an antiderivative of $f(x)$.

Draw the segment from P to (a, y_1) and a parallel segment from $[a, F(a)]$. Both segments are labeled (1) in Figure 10.8. The latter segment intersects $x = x_1$ at $F(x_1)$. Since the slopes of these parallel segments are equal,

$$\frac{y_1}{1} = \frac{F(x_1) - F(a)}{x_1 - a}$$

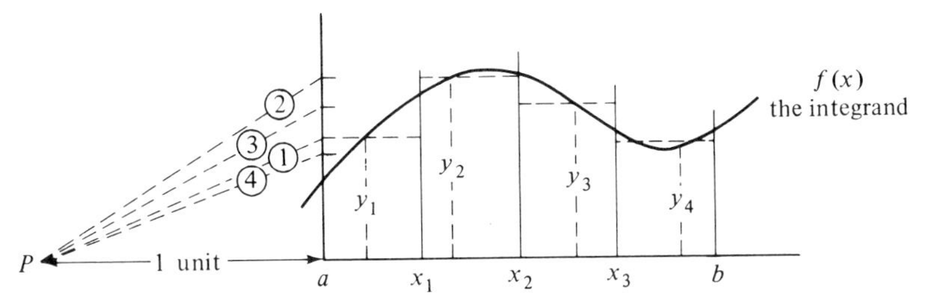

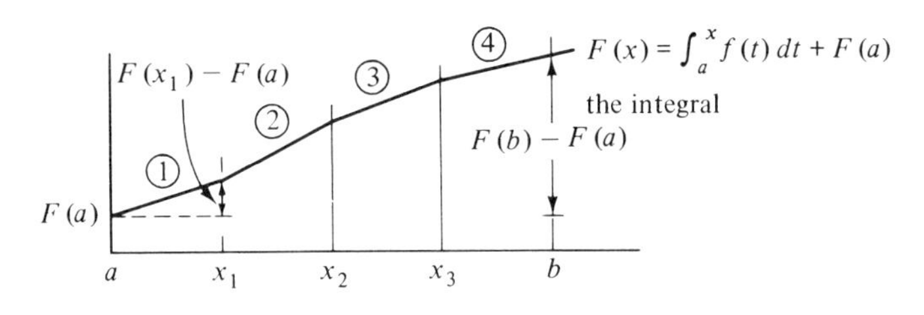

Fig. 10.8

$$F(x_1) - F(a) = y_1(x_1 - a) \simeq \int_a^{x_1} f(x)\,dx$$

For the subsequent panels,

$$\begin{aligned} F(x_2) - F(x_1) &= y_2(x_2 - x_1) \simeq \int_{x_1}^{x_2} f(x)\,dx \\ &\vdots \\ F(b) - F(x_{n-1}) &= \qquad\qquad \simeq \int_{x_{n-1}}^{b} f(x)\,dx \end{aligned}$$

Adding,

$$F(b) - F(a) \simeq \int_a^b f(x)\,dx$$

10.1.6 Charge-a-Condenser. Given a condenser of known capacitance C, a voltmeter, and a device to produce a current whose variation with time is $f(t)$ (Fig. 10.9). If the voltage is $V(a)$ at time $t = a$, the voltage at time $t = b$ determines the value of the integral

$$\int_a^b f(t)\,dt = C[V(b) - V(A)]$$

$i = f(t)$ C V

Fig. 10.9

Such a device is an *electrical analog* that may be used to approximate the value of an integral. Electrical analog computers are manufactured in which appropriate electrical units (such as adders, multipliers, and integrators) may be joined to represent a given mathematical statement. These are *not* digital computers but play an important role in the world of computational machinery.

10.1.7 A Riemann Sum. The integral $\int_a^b f(x)\,dx$ is defined as the limit of a certain sequence of sums (Riemann). These sums are formed in the following way:

1. Partition $[a, b]$ into $[x_0, x_1, x_2, \ldots, x_n]$, where $x_0 = a$ and $x_n = b$.
2. Select in each subinterval of length $\Delta x_i = x_i - x_{i-1}$ a value of x, say ξ_i.
3. The corresponding Riemann sum is $\sum_{i=1}^{n} f(\xi_i)\,\Delta x_i$.

The definite integral is the limit of all sequences of such sums for which $n \longrightarrow \infty$ and $\Delta x_i \longrightarrow 0$, if such a limit exists.

$$\int_a^b f(x)\,dx = \lim_{\substack{n\to\infty \\ \Delta x_i \to 0}} \sum_{i=1}^{n} f(\xi_i) x_i$$

An approximate value of the integral is an element of such a sequence. For example,

$$\int_0^1 \frac{dx}{1+x^2}, \qquad \text{partition: } [0, \tfrac{1}{4}, \tfrac{1}{2}, \tfrac{3}{4}, 1], \quad \xi_i\text{: midpoints}$$

ξ_i	$f(\xi_i)$
$\frac{1}{8}$	$\frac{64}{65} \simeq .985$
$\frac{3}{8}$	$\frac{64}{73} \simeq .877$
$\frac{5}{8}$	$\frac{64}{89} \simeq .720$
$\frac{7}{8}$	$\frac{64}{113} \simeq .566$
	$\sum f(\xi_i) = 3.148$

$$\int_0^1 \frac{dx}{1+x^2} \simeq \sum_{i=1}^{4} f(\xi_i)\,\Delta x_i = (3.148)(\tfrac{1}{4}) = .787$$

The definite integral was also defined in Section 4.20, and examples were proposed to show that

1. for an integrable function, the value of the definite integral over $[a, b]$ is independent of the choice of ξ_i in each subinterval of $[a, b]$.
2. The limit requires that both $n \longrightarrow \infty$ and $\Delta x_i \longrightarrow 0$.

10.1.8 Integrand Approximated by a Taylor Polynomial. If the integrand $f(x)$ can be approximated in $[a, b]$ as a polynomial in $x - c$, where $a < c < b$ in the Taylor sense, then

$$\begin{aligned}\int_a^b f(x)\,dx &= \int_a^b [P_n(x - c) + R(x)]\,dx \\ &= \int_a^b P_n(x - c)\,dx + \int_a^b R(x)\,dx \\ &= \int_a^b \left[f(c) + f'(c)(x - c) + \cdots + \frac{f^{(n)}(c)}{n!}(x - c)^n \right] dx + E\end{aligned}$$

The fundamental theorem of calculus can be employed to evaluate $\int_n^b P_n(x - c)\,dx$, since the antiderivative of a polynomial is known; for example,

$$\begin{aligned}\int_0^{.1} e^{-x^2}\,dx &= \int_0^{.1} \left(1 - x^2 + \frac{x^4}{2!}\right) dx + E \\ &= .09966767 + E\end{aligned}$$

From Taylor's theorem it can be shown that for this example

$$|E| < \tfrac{1}{42} \cdot 10^{-7} \simeq .5 \times 10^{-8}$$

An intuitive recognition that this particular approximation is a good one (the error is small) can be had by sketching the graphs of e^{-x^2} and its approximating polynomial $1 - x^2 + (x^4/2)$, (Fig. 10.10). If the upper limit had been 1., the error obviously would be large.

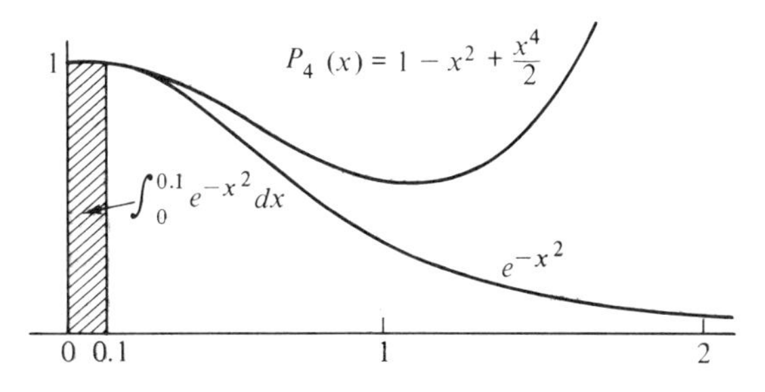

Fig. 10.10

10.2 Newton–Cotes Quadrature Formulas

If integrand $f(x)$ can be approximated in $[a, b]$ by a polynomial of degree n interpolating $f(x)$ at $n + 1$ points,

$$\int_a^b f(x)\,dx = \int_a^b (P_n(x) + R(x))\,dx$$

$$\int_a^b f(x)\,dx = \int_a^b P_n(x)\,dx + \int_a^b R(x)\,dx$$

$$= Q_n + E$$

The quadrature formula Q_n is an approximation of the integral and E the error. In the remaining portion of this chapter the construction and use of some quadrature formulas will be considered, but again the study of error will be deferred to a course in numerical analysis.

The Newton–Cotes quadrature formulas are those resulting from

1. $P_n(x)$ interpolating $f(x)$ at *equally spaced points*, that is, $x_i - x_{i-1} = h$, where $i = 1, 2, \ldots, n$.

2. The limits of integration are elements of the set $x_0 + mh$ and are *symmetric* with respect to x_0 and x_n.

If $a = x_0$ and $b = x_n$ the integration is over n panels and the quadrature formula Q_n is said to be *closed* (Fig. 10.11).

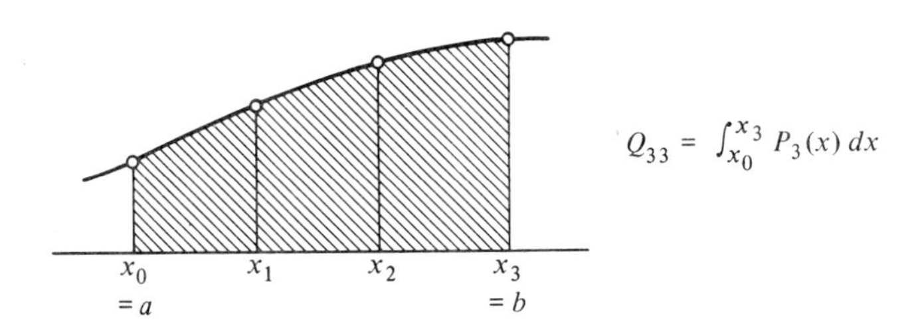

Fig. 10.11

If $a = x_0 + mh$, $b = x_n - mh$, and $a < b, m > 0$, integration takes place over $n - 2m$ panels and the quadrature formula Q_n is said to be *open* (Fig. 10.12).

If $a = x_0 - mh, b = x_n + mh$, and $a < b, m > 0$, integration takes place over $n + 2m$ panels and the quadrature formula Q_n is said to be an *extrapolating* formula (Fig. 10.13).

For the notation Q_{np}, let n be the degree of the interpolating polynomial (to $n + 1$ equally spaced points) and p the number of symmetrically placed panels over which integration takes place.

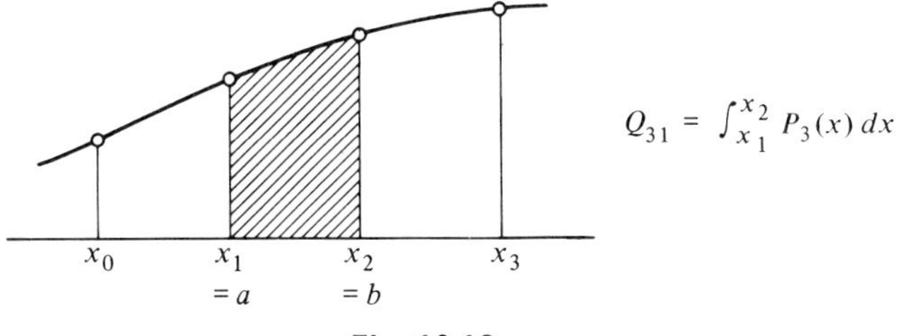

Fig. 10.12

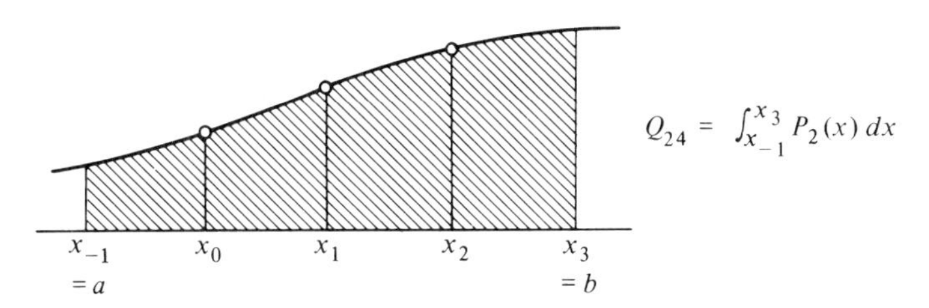

Fig. 10.13

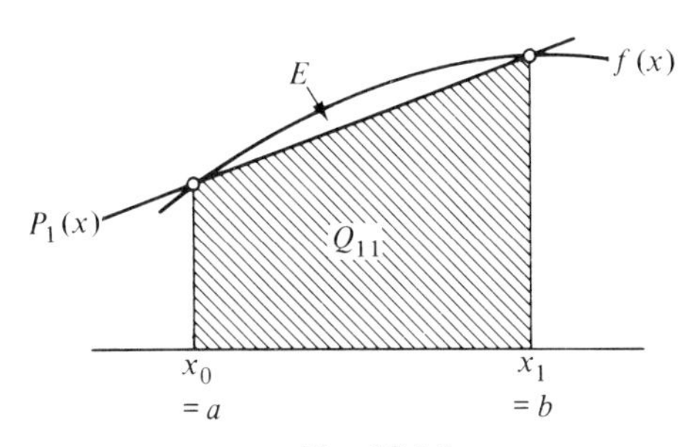

Fig. 10.14

To derive Q_{11} (Fig. 10.14),

$$\int_{x_0}^{x_1} f(x)\,dx = \int_{x_0}^{x_1} P_1(x)\,dx + \int_{x_0}^{x_1} R(x)\,dx$$

Expressing $P_1(x)$ in Lagrange form,

$$\begin{aligned}\int_{x_0}^{x_1} f(x)\,dx &= \int_{x_0}^{x_1} \left(\frac{x - x_1}{x_0 - x_1} y_0 + \frac{x - x_0}{x_1 - x_0} y_1\right) dx + E \\ &= \frac{1}{2h}[-(x - x_1)^2 y_0 + (x - x_0)^2 y_1]\Big|_{x = x_0}^{x = x_1} + E \\ &= \frac{h}{2}(y_0 + y_1) + E\end{aligned}$$

thus

$$Q_{11} = \frac{h}{2}(y_0 + y_1)$$

For example,

$$\int_0^1 \frac{dx}{1+x^2} = \tfrac{1}{2}(1 + \tfrac{1}{2}) + E = .75 + E$$

To derive Q_{22} (Fig. 10.15),

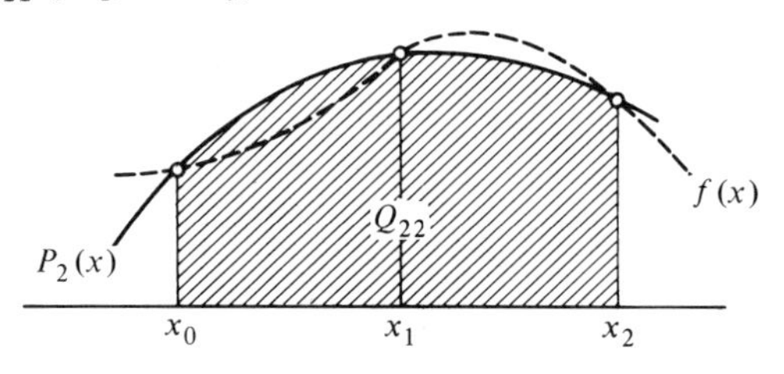

Fig. 10.15

$$\int_{x_0}^{x_2} f(x)\,dx = \int_{x_0}^{x_2} P_2(x)\,dx + \int_{x_0}^{x_2} R(x)\,dx$$

$$= \int_{x_0}^{x_2}\left[\frac{(x-x_1)(x-x_2)}{(x_0-x_1)(x_0-x_2)}y_0 + \frac{(x-x_0)(x-x_2)}{(x_1-x_0)(x_1-x_2)}y_1 + \frac{(x-x_0)(x-x_1)}{(x_2-x_0)(x_2-x_1)}y_2\right]dx + E$$

This awkward integration can be simplified thanks to the "equal spacing" of data, which suggests the linear transformation

$$x = x_0 + sh, \quad dx = h\,ds$$

$$\int_{x_0}^{x_2} f(x) = h\left[y_0 \int_0^2 \tfrac{1}{2}(s-1)(s-2)\,ds - y_1 \int_0^2 s(s-2)\,ds + y_2 \int_0^2 \tfrac{1}{2}s(s-1)\,ds\right] + E$$

$$= h(A_0 y_0 + A_1 y_1 + A_2 y_2) + E$$

The coefficients A_0, A_1, A_2 are determined by n, the degree of the interpolating polynomials, and p, the number of panels. It is important to observe that A_0, A_1, A_2 are independent of h and the abscissas x_0, x_1, x_2. It is suggested that the student perform these integrations so that he may write with conviction:

$$Q_{22} = \frac{h}{3}(y_0 + 4y_1 + y_2]$$

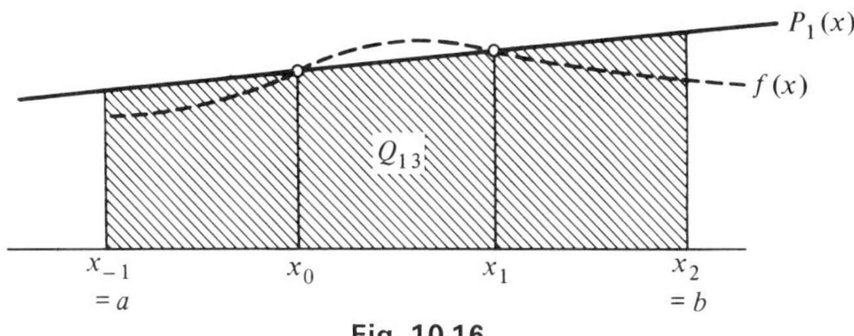

Fig. 10.16

To derive Q_{13} (Fig. 10.16), an extrapolating quadrature formula:

$$\int_a^b f(x)\,dx = \int_{x_{-1}}^{x_2} \left(\frac{x - x_1}{x_0 - x_1} y_0 + \frac{x - x_0}{x_1 - x_0} y_1\right) dx + E$$

Again, let

$$x = x_0 + sh$$

$$= h \int_{-1}^{2} [-(s - 1)y_0 + sy_1]\,ds + E$$

$$= \frac{3h}{2}(y_0 + y_1) + E$$

Tabulated below are some of the lower-ordered Newton–Cotes quadrature formulas,

$$Q_{np} = h \sum_{i=0}^{n} A_i y_i$$

(A more extensive tabulation can be found in Kunz under "Numerical Analysis" in References.)

	A_0	A_1	A_2	A_3
Q_{11}	$\frac{1}{2}$	$\frac{1}{2}$		
Q_{13}	$\frac{3}{2}$	$\frac{3}{2}$		
Q_{22}	$\frac{1}{3}$	$\frac{4}{3}$	$\frac{1}{3}$	
Q_{24}	$\frac{8}{3}$	$-\frac{4}{3}$	$\frac{8}{3}$	
Q_{31}				
Q_{33}	$\frac{3}{8}$	$\frac{9}{8}$	$\frac{9}{8}$	$\frac{3}{8}$
Q_{35}	$\frac{55}{24}$	$\frac{5}{24}$	$\frac{5}{24}$	$\frac{55}{24}$

10.2.1 Fill in the missing row in the above table for Q_{31}.

10.2.2 Approximate

$$\int_0^1 \frac{dx}{1+x^2} = \frac{\pi}{4} = .785398\ldots$$

by Q_{11}:

$$\int_0^1 \frac{dx}{1+x^2} = .75 + E, \qquad E = .0354\ldots$$

by Q_{22}:

$$\int_0^1 \frac{dx}{1+x^2} = .783 + E, \qquad E = .0024\ldots$$

by Q_{33}:

$$\int_0^1 \frac{dx}{1+x^2} = .7846 + E, \qquad E = .0008\ldots$$

The calculations for Q_{22} and Q_{33} are as follows:

x_i	y_i	A_i	$A_i y_i$
.0	1.0	1/3	1.0/3
.5	.8	4/3	3.2/3
1.0	.5	1/3	.5/3
			4.7/3

$Q_{22} = h*1.567$

x_i	y_i	A_i	$A_i y_i$
.0	1.0	3/8	1.0*3/8
.333 . . .	.9	9/8	2.7*3/8
.666 . . .	.692	9/8	2.076*3/8
1.0	.5	3/8	.5*3/8

$Q_{33} = h*2.354$

Some very useful quadrature formulas are derived from a repeated usage of the Newton–Cotes formulas.

The *trapezoidal rule* is an algorithm for the approximation of

$$\int_{x_0}^{x_n} f(x)\,dx$$

if $f(x_k)$ is given for the equally spaced values $x_k = x_0 + kh$, where $k = 0, 1, 2, \ldots, n$. It is formed by the repeated use of Q_{11}.

$$\begin{aligned}\int_{x_0}^{x_n} f(x)\,dx &= \int_{x_0}^{x_1} f(x)\,dx + \int_{x_1}^{x_2} f(x)\,dx + \cdots + \int_{x_n}^{x_{n-1}} f(x)\,dx \\ &= \frac{h}{2}(y_0 + y_1) + E_1 + \frac{h}{2}(y_1 + y_2) + E_2 + \cdots \\ &\quad + \frac{h}{2}(y_{n-1} + y_n) + E_n\end{aligned}$$

$$= \frac{h}{2}(y_0 + 2y_1 + 2y_2 + \cdots + 2y_{n-1} + y_n) + E,$$

$$E = E_1 + E_2 + \cdots + E_n$$

10.2.3 For an even number of panels (an odd number of points) derive *Simpson's rule*,

$$\int_{x_0}^{x_n} f(x)\,dx = \frac{h}{3}(y_0 + 4y_1 + 2y_2 + 4y_3 + \cdots + 2y_{n-2} + 4y_{n-1} + y_n) + E$$

through the repeated usage of Q_{22}.

As an example, write a program to approximate the area of the figure bounded by

$$r = (\cos\theta)^{1/3}$$

Use the trapezoidal Rule and $n = 100$.

Figure 10.17 shows the graph of $r^3 = \cos\theta$ in the polar plane. Because of symmetry of the bounded region with respect to the polar axis,

$$\text{area} = 2\int_{\theta=0}^{\theta=\pi/2} dA = 2\int_0^{\pi/2} \tfrac{1}{2}r^2\,d\theta = \int_0^{\pi/2} (\cos\theta)^{2/3}\,d\theta$$

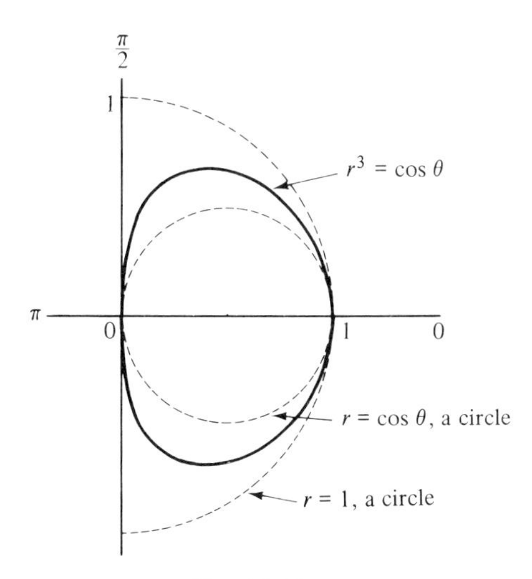

Fig. 10.17

```
C     AREA BOUNDED BY R**3=COS(THETA)
      F(T)=(COS(T))**(2./3.)
      A=0.
      B=3.141593/2.
      N=100
      AN=N
      H=(B-A)/AN
      Y0=F(A)
      YN=F(B)
      SUM=Y0+YN
      M=N-1
      DO 1 I=1,M,1
        AI=I
      X=AI*H
    1 SUM=SUM+2.*F(X)
      AREA=H*SUM/2.
      WRITE (6,100) AREA
      STOP
  100 FORMAT('b',F12.7)
      END
```

10.2.4 Write a program to approximate $\int_a^b f(x)\,dx$ using Simpson's rule, with a statement subroutine defining the integrand and statements to define A, B, and N.

Appropriate to each of the following problems sketch the figure and write a statement subroutine for the integrand and statements for A, B, and N.

10.2.5 Find the area of the figure bounded by

$$x^4 + y^4 = 1, \quad h = .1.$$

ANSWER:

```
F(X)=4.*(1.-X**4)**.25
A=0.
B=1.
N=10
```

Find the area of the plane figure bounded by

10.2.6 $$r = 1 + \tan\theta, \quad \theta = \frac{\pi}{4}, \quad h = \frac{\pi}{20}$$

10.2.7 $$y = 0, \quad \begin{cases} x = \cos t, \\ y = \sin^2(t+1), \end{cases} \quad 0 \leqq t \leqq \pi, \quad h = \Delta t = \frac{\pi}{20}$$

10.2.8 $x^4 - y^4 = 1, \quad x = 2, \quad h = .01$

10.2.9 inside $r = 9 \cos^3 \theta$, outside $r = 1 + \cos^3 \theta$

Find the length of arc of

10.2.10 $y = x^2, \quad 0 \leqq x \leqq 1, \quad h = .01$

10.2.11 $\dfrac{x^2}{2^2} + \dfrac{y^2}{3^2} = 1, \quad h = .01$

10.2.12 $r^3 = \cos \theta, \quad h = \Delta\theta = \dfrac{\pi}{200}$

10.2.13 $x = \sin t, \quad y = \log (t + 1), \quad 0 \leqq t \leqq \dfrac{\pi}{2}, \quad h = \dfrac{\pi}{40}$

10.2.14 $r = 2 + \sec \theta, \quad \dfrac{2\pi}{3} < \theta < \pi, \quad h = \dfrac{\pi}{30}$

Find the volume generated by rotating about the x-axis the plane figure bounded by

10.2.15 $y^4 = x(1 - x), \quad 0 \leqq x \leqq 1, \quad h = .1$

10.2.16 $x = \sin t, \quad y = t\left(t - \dfrac{\pi}{2}\right)(t - \pi), \quad 0 \leqq t \leqq \pi, \quad h = \dfrac{\pi}{20}$

10.2.17 $r^3 = \cos \theta, \quad h = \dfrac{\pi}{20}$

For the following problems (10.2.18–10.2.20) write a program to print out the centroid of the defined plane figure. When possible use the same program, changing only statement subroutines, limits, and increment. For these problems, the use of a subroutine for integration would be an asset. (Consult Anderson under "Programming" in References.)

10.2.18 $y^4 = x(1 - x), \quad h = .1$

10.2.19 $x = \sin t, \quad y = t\left(t - \dfrac{\pi}{2}\right)(t - \pi), \quad 0 \leqq t \leqq \pi, \quad \Delta t = \dfrac{\pi}{20}$

10.2.20 $$r^3 = \cos\theta, \quad h = \frac{\pi}{20}$$

The following two problems require a sequence of calculations.

10.2.21 Find the areas of the figures in the first quadrant bounded by

$$x^n + y^n = 1, \quad 0 \leqq x \leqq 1, \qquad n = 1, 2, 3, \ldots, 10$$

10.2.22 Find the length of arc in the first quadrant of

$$x^n + y^n = 1, \quad 0 \leqq x \leqq 1, \qquad n = 1, 2, 1, \ldots, 10$$

(To avoid the noncomputability of dy/dx at $x = 1$, make use of the symmetry of these figures with respect to $y = x$. Use $N = 100$ for each integral.)

10.2.23 Find the area bounded by each of the following:

$$r = (\cos\theta)^k, \quad k = 8, 4, 2, 1, \tfrac{1}{2}, \tfrac{1}{4}, \tfrac{1}{8}, \quad |\theta| \leqq \frac{\pi}{2}$$

Find the area of the figure bounded by the following curves. In each case it is necessary to find the intersection of two curves, before the integral representing the area can be established.

10.2.24 $$y = \frac{e^{-x}}{x+1}, \quad y = x, \quad y = 0, \quad N = 100$$

10.2.25 $$y = x \log x, \quad y = 2 - \frac{x^2}{2}, \quad y = 0, \quad N = 100$$

Each of the following problems (10.2.26–10.2.30) differ from the preceding in that the integrand is defined in terms of a set of number pairs read in as data. In each case construct an appropriate program.

10.2.26 A water tank of axial symmetry is defined by a set of 21 diameters $X(I)$, measured at each foot of elevation in the tank, where $y = 0, 1, 2, \ldots,$ 20 ft.

Let the 21 values of X be punched one to a card, each in F10.4. Write a program to read in these diameters and print out the volume of water as a function of depth. Use the trapezoidal rule.

10.2.27 An amphora of axial symmetry (Fig. 10.18) has its cross section defined by

$$X(I),\ Y(I); \qquad Y(41) - Y(1) = 4 \text{ ft}$$
$$Y(I+1) - Y(I) \qquad = .1 \text{ ft}$$

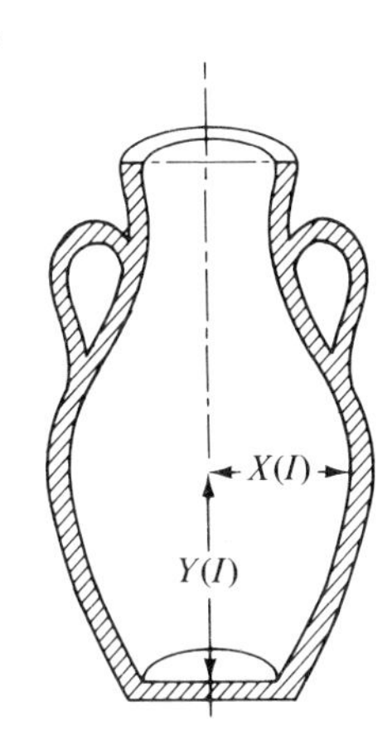

Fig. 10.18

The vessel is filled to a depth of 3 ft with water. Use Simpson's rule to approximate the work necessary to pump the water over the rim of the amphora.

10.2.28 A dam 100 ft high is built across a canyon (Fig. 10.19) whose section is defined by

$$X(I),\ Y(I); \qquad Y(I) = (I-1)\,ft, \quad I = 1, 2, \ldots, 101$$

Find approximate values for the force on the dam and the elevation of the *center of pressure*.

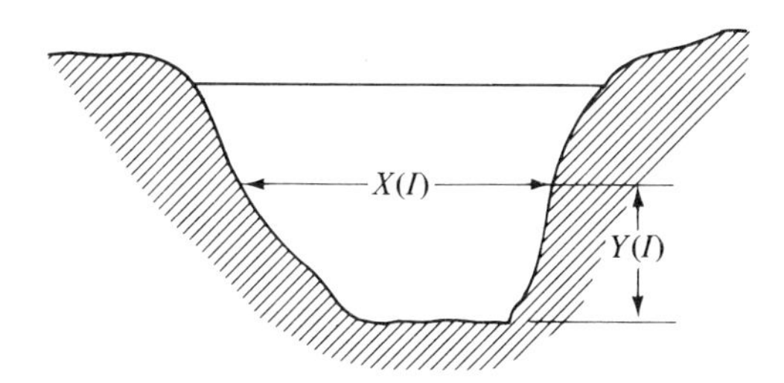

Fig. 10.19

10.2.29 The cross section of a railway rail (Fig. 10.20) is defined by $X(I)$, $Y(I)$ where $I = 1, 2, \ldots, 10$, with one pair of data punched per card, and each number in F10.4.

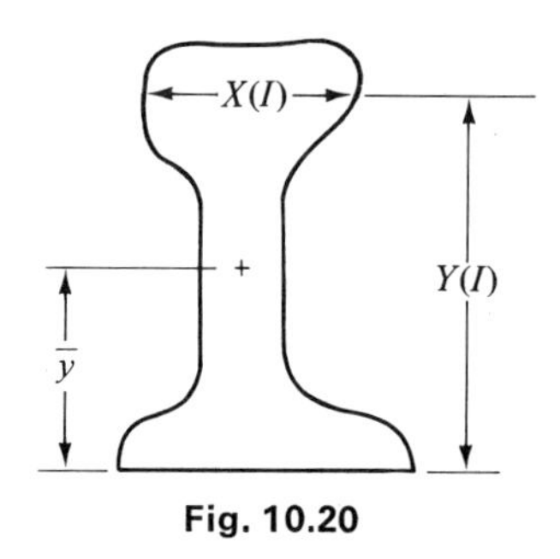

Fig. 10.20

The ordinate of the centroid of the section is

$$\bar{y} = \frac{\int_A y \cdot x \, dy}{\int_A x \, dy}$$

The *moment of inertia* (a measure of "stiffness") with respect to a horizontal line through the centroid is

$$\begin{aligned} S &= \int_A (y - \bar{y})^2 x \, dy \\ &= \int_A y^2 x \, dy - 2\bar{y} \int_A yx \, dy + \bar{y}^2 \int_A x \, dy \\ &= \int_A y^2 x \, dy - \bar{y}^2 \int_A x \, dy \end{aligned}$$

Read in the data defining the section and print out the area of the section, $\bar{y}$, and S.

10.2.30 The cross section of a beam is in the shape of an air foil (Fig. 10.21) The perimeter is defined by the coordinate pairs

$X(I)$, $B(I)$ bottom curve

$X(I)$, $T(I)$ top curve

where

$$X(I) = (I - 1)\frac{L}{N}, \qquad I = 1, 2, \ldots, N + 1$$

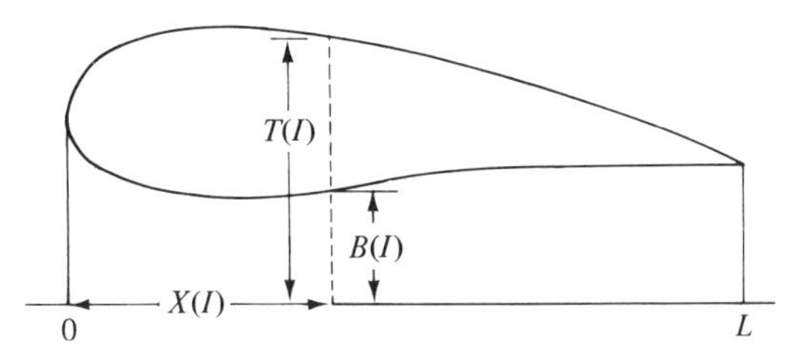

Fig. 10.21

Write a program to read in L, N, $B(I)$, $T(I)$, where $I = 1, 2, \ldots, N + 1$ and print out the following:

(a) The area of the figure.

(b) The abscissa of its centroid.

(c) The moment of inertia about a vertical line through the centroid. Use the trapezoidal rule.

10.2.31 Following is a "cut-and-fill" problem. A road of constant width is to be built across a valley and mountain, as shown in Figure 10.22. A survey provides $N + 1$ equal interval stations and the corresponding elevations $Y(I)$. Before grading begins, we can approximate the net area (mountain minus valley) by, for instance, the trapezoidal rule. We would like to grade so that the total dirt cut is equal to the fill and also have as the resulting grade

$$y(x) = bx^2(x - c)^2$$

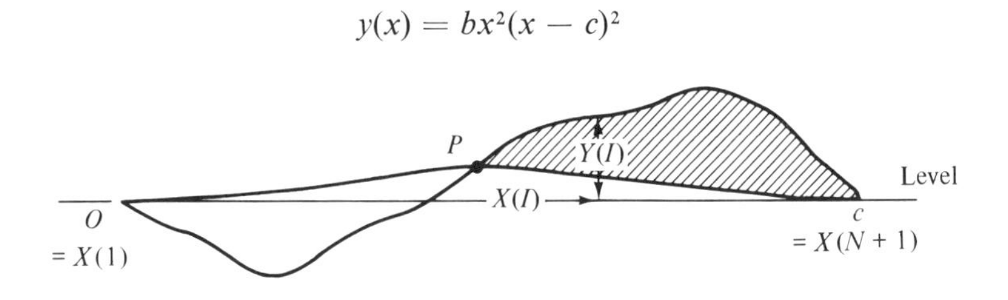

Fig. 10.22

(a) Write a program to find the net area A (mountain minus valley).

(b) Find the coefficient b if $A = \int_o^c bx^2(x - c)^2 \, dx$.

(c) Print out the station nearest and to the right of point P.

(d) Print out the amount of earth to be cut (the shaded region).

(e) Print out elevations for the new road at each station.

(f) For input data let C be 2,000 feet, $N = 20$, and the maximum elevation of the mountain 300 feet. Scale from the figure the values of $Y(I)$.

10.2.32 Develop a quadrature formula over a portion of n equal intervals, n divisible by 3, using Q_{33} repeatedly. Using this formula, evaluate

$$\int_0^1 \frac{dx}{1+x^2}, \qquad n = 18$$

10.2.33 An entirely different method of approximating an integral was proposed in Section 10.1.3. The Monte Carlo method requires that for each trial there be generated two numbers, θ and ϕ, each in [0, 1] with random digits. The following is a suggestion for generating random numbers (see Maisel under "Programming" in References). These numbers may fall short of satisfying various tests designed to verify randomness, but they will suffice for our purpose.

Select four (or more) random numbers, each with a number of digits equal to the computer decimal word length. For instance, from Abramowitz and Stegun, Table 26.11,

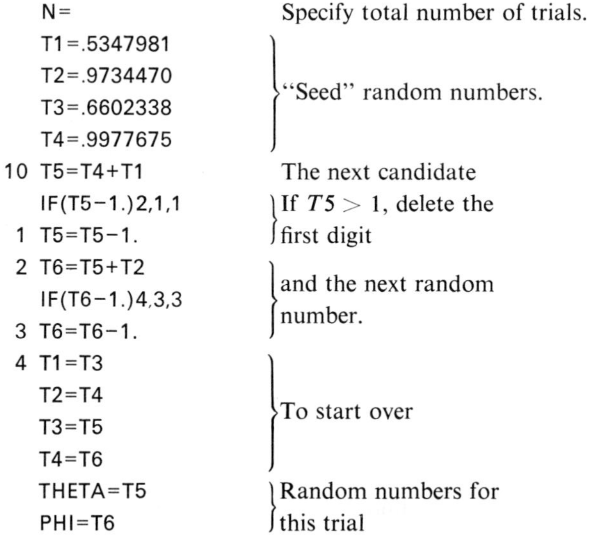

```
      N=                   Specify total number of trials.
      T1=.5347981        }
      T2=.9734470        }
      T3=.6602338        } "Seed" random numbers.
      T4=.9977675        }
   10 T5=T4+T1             The next candidate
      IF(T5-1.)2,1,1     } If T5 > 1, delete the
    1 T5=T5-1.           } first digit
    2 T6=T5+T2           }
      IF(T6-1.)4,3,3     } and the next random
    3 T6=T6-1.           } number.
    4 T1=T3              }
      T2=T4              }
      T3=T5              } To start over
      T4=T6              }
      THETA=T5           } Random numbers for
      PHI=T6             } this trial
```

In this portion of the program the total number trials and the number of hits are recorded (see 10.1.3). Test the trial number against N, if another trial is to be taken,

```
      GO TO 10
```

Given π, write a program to approximate

$$\frac{1}{\sqrt{2\pi}} \int_0^1 e^{-(x^2/2)}\, dx$$

by 10, 100, and 1000 trials in the Monte Carlo procedure. This approximation (given in Abramowitz and Stegun, p. 966) is .34134 47460 The integrand is called the standard normal probability density function.

This is not a very efficient procedure to approximate this integral to a prescribed number of decimal places. In a study of probability it can be shown that the expected deviation of this result is $\sqrt{.16/n}$. Thus for $n = 10{,}000$ trials we may expect only two significant digits. To find the next digit, 10^6 trials would be needed.

10.2.34 Approximate by the Monte Carlo method for $N = 10, 100, 1000$ trials,

$$\int_0^1 \frac{dx}{1+x}$$

and check these results against a tabulated value of this integral.

10.2.35 Determine an approximate value of the integral $\int_a^b f(x)\,dx$ in the following way:

1. Select a positive integer N.
2. Generate N random numbers θ_i in [0, 1], and for each compute $f(x_i)$, where

$$x_i = a + (b - a)\theta_i$$

3. Let $\Delta x = (b - a)/N$.
4. Compute $\sum_{i=1}^N f(x_i)\,\Delta x = \Delta x \sum_{i=1}^N f(x_i) \approx \int_a^b f(x)\,dx$. Approximate

$$\frac{1}{\sqrt{2\pi}} \int_0^1 e^{-(x^2/2)}\,dx$$

by this device for $N = 10, 100, 1000$.

10.3 Functions in Integral Form

Given $f(x)$ we may define a corresponding function,

$$F(x) = \int_a^x f(t)\,dt$$

called the integral of $f(x)$. For a specified x, the value of $F(x)$ can be interpreted as the area of the figure bounded by $y = f(t)$, $y = 0$, $t = a$, $t = x$ (Fig. 10.23).

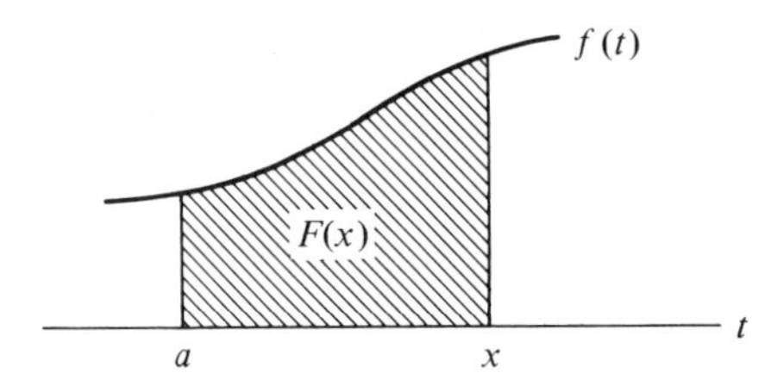

Fig. 10.23

From the definition of the definite integral,

$$F(a) = 0$$

and the fundamental theorem of calculus states that

$$\frac{dF(x)}{dx} = f(x)$$

Some examples follow.

In some calculus textbooks the logarithm function is defined as

$$\log(x) = \int_1^x \frac{1}{t} dt$$

The sine–integral function (see Abramowitz and Stegun, p. 228) is defined as

$$\text{Si}(x) = \int_0^x \frac{\sin t}{t} dt$$

An exponental–integral function (see Abramowitz and Stegun, p. 228) is defined in terms of an improper integral,

$$E_1(x) = \int_x^\infty \frac{e^{-t}}{t} dt$$

Some functions are defined in terms of an integral where the independent variable appears only in the integrand,

$$F(x) = \int_a^b f(x, t)\, dt$$

For a specified value of x, say $x = x_1$, $F(x_1)$ can be interpreted as the area of the figure bounded by

$$y = f(x_1, t), \quad y = 0, \quad t = a, \quad t = b$$

A change in parameter x, in general, causes a variation in the upper boundary of the figure and thus a change in $F(x)$, the area under the curve. For example, the "complete elliptic integral of the first kind" (see Abramowitz and Stegun, p. 608) is as follows:

$$F(m) = \int_0^{\pi/2} (1 - m \sin^2 \theta)^{-1/2} \, d\theta = \int_0^1 \frac{dt}{\sqrt{(1 - t^2)(1 - mt^2)}}, \qquad |m| < |$$

and "the complete elliptic integral of the second kind" is

$$G(m) = \int_0^{\pi/2} (1 - m \sin^2 \theta)^{+1/2} \, d\theta = \int_0^1 \sqrt{\frac{1 - mt^2}{1 - t^2}} \, dt$$

It is interesting from a computational point of view that the trigonometric form of the above functions is a proper integral but that the substitution $\sin \theta = t$ produces an equivalent form that is an improper integral. Why?

The Bessell function of zero order and first kind (see Abramowitz and Stegun, p. 355) is

$$J_0(x) = \frac{1}{\pi} \int_0^{\pi} \cos(x \sin t) \, dt$$

The gamma function is defined in terms of an improper integral (see Abramowitz and Stegun, p. 253)

$$\Gamma(x) = \int_0^{\infty} t^{x-1} e^{-t} \, dt$$

10.3.1 Write a program to tabulate

$$\log (1 + x) = \int_0^x \frac{dt}{1 + t}$$

for $x = [0, .01, 1]$, using the trapezoidal rule. Your results may be compared with those tabulated in Abramowitz and Stegun, page 100.

```
      X=0.
      AREA=0.
      WRITE (6,100) X,AREA
      DO 1 I=1,100,1
        X1=X+.01
        AREA=AREA+.005*(1./(1.+X)+1./(1.+X1))
        WRITE (6,100) X,AREA
    1 X=X1
      STOP
  100 FORMAT('b',F5.2,F20.7)
      END
```

10.3.2 Write a program to tabulate

$$\text{Si}(x) = \int_0^k \frac{\sin t}{t}\,dt$$

for $x = [0, .01, 1.]$, using the trapezoidal rule. The results for $.5 \leq x \leq 1.0$ can be compared with those tabulated in Abramowitz and Stegun, page 239. Note that this problem, except for the integrand, is identical to the above example. Since this integrand is not computable at $t = 0$, the integration for the first panel must be done before the DO statement.

10.3.3 Tabulate the functions $E(t)$ and $W(t)$ for $t = n(.1)$, where $n = 0, 1, 2, \ldots, 30$, if

$$W(t) = \int_0^t P(z)\,dz, \quad P = \frac{E^2}{10.}, \quad E = e^{-t} \sin t$$

10.3.4 Tabulate the function

$$F(x) = \frac{1}{\sqrt{2\pi}} \int_0^x e^{-(t^2/2)}\,dt$$

for $x = n(.01)$, where $n = 0, 1, 2, \ldots, 30$,

(a) Using Q_{11} and $h = .01$.

(b) Using Q_{22} and $h = .005$.

Compare your results with a tabulation of the standard normal distribution function.

10.3.5 Write a program to tabulate

$$E_1(x) = \int_x^{\infty} \frac{e^{-t}}{t}\, dt$$

for $x = [1, .1, 2.]$, using the trapezoidal rule. First compute $E_1(2.)$. Let $h = .1$ for $2. \leqq t \leqq 4.$, then for $t > 4$. Let $h = .5$ until $e^{-t}/t < 10^{-5}$. Next, compute $E_1(1.9)$, $E_1(1.8)$, and so forth, saving each in memory. Then print out the tabulation of x and $E_1(x)$.

This is an arbitrary device to cope with the problem of integration to infinity when we are using a machine that can perform only a finite number of operations. This device may have intuitive appeal but leaves the problem of error estimation unresolved. Compare your results to those tabulated in Abramowitz and Stegun, page 239.

10.3.6 Write a program to tabulate

$$F(x) = \int_0^1 \frac{dt}{x + t}$$

for $x = [1., .1, 2.]$, using $h = .01$ and the trapezoidal rule.

In this example the antiderivative is known, thus

$$F(x) = \log\left(\frac{x + 1}{x}\right)$$

Considering the Fortran values of ALOG to be exact, we could tabulate $F(X)$ computed through the quadrature formulas, F(X) = ALOG((X + 1.)/X), and the error.

```
  X=.9
  DO 1 I=1,11,1
    X=X+.1
    FXQ=1./X+1./(X+1.)
    T=0.
    DO 2 J=1,99,1
      T=T+.01
2   FXQ=FXQ+2.*(1./(X+T))
    FXQ=.005*FXQ
    FX=ALOG((X+1.) X)
```

```
      ER=FX-FXQ
    1 WRITE (6,100) X,FXQ,FX,ER
  100 FORMAT('b',4F12,7)
      STOP
      END
```

10.3.7 Tabulate $f(x) = \int_1^2 \log(xt)\,dt$ for $x = [1., .1, 3.]$, using $h = .1$ and the trapezoidal rule. Reconsider the integrand as $\log x + \log t$, and use the fundamental theorem to check your computed results.

10.3.8 Tabulate $J_0(x) = 1/\pi \int_0^\pi \cos(x \sin t)\,dt$ for $x = [0., .1, 1.]$, using $h = .01$ and Simpson's rule. Compare your results with those tabulated in Abramowitz and Stegun, page 390.

10.3.9 Tabulate $g(x) = \int_0^\infty e^{-xt^4}\,dt$ for $x = [1., .1, 2.]$. Use the trapezoidal rule and $h = .1$ until $e^{-xt^4} < 10^{-5}$.

10.3.10 Tabulate $\Gamma(x) = \int_0^\infty t^{x-1}e^{-t}\,dt$ for $x = [1., .1, 2.]$. Let $h = .1$ for $t \leqq 4.$ and $h = 1.$ for $t > 4.$, and continue the integration for each tabular entry until the integrand is less than 10^{-5}. Compare your print out with values tabulated in Abramowitz and Stegun, page 267.

By use of the device of integration by parts it can be shown that

$$\Gamma(x + 1) = x\Gamma(x)$$

Using this formula continue the tabulation for $x = [2.1, .1, 4.]$.

10.3.11 Write a program to read in *FO*, and *Y(I)*, where $I = 1, 2, 3, \ldots,$ 101, values of $y(t)$ for $t = [0, .01, 1.]$. Print out a tabulation of the function

$$F(x) = FO + \int_0^x y(t)\,dt$$

for $x = [0, .1, 1.]$. Use the trapezoidal rule.

10.3.12 Write programs to tabulate the following functions for $x = [0., .1, 1.]$, using the trapezoidal rule:

(a) $$F(x) = \int_0^x \sin\frac{1}{t+1}\,dt$$

(b) $$G(x) = \int_0^1 \sin\frac{x}{t+1}\,dt$$

(c) $$H(x) = \int_0^x \sin\frac{x}{t+1}\,dt.$$

10.4 Initial Value Problems

The problem is to determine the function $F(x)$ corresponding to the given $f(x)$, x_0, and F_0 such that

$$\frac{dF(x)}{dx} = f(x), \quad F(x_0) = F_0$$

From the fundamental theorem of calculus, an equivalent definition of $F(x)$ is

$$F(x) = F_0 + \int_{x_0}^{x} f(t)\,dt$$

10.4.1 A vessel is being filled with water. At time t, the volume is $V(t)$. If it is known that

$$\frac{dV}{dt} = \sqrt{1+\sqrt{t}}, \quad V(0) = 0$$

tabulate $V(t)$ for $t = [0., .1, 1.]$. If the vessel being filled is a cone with vertex angle 90°, also tabulate depth as a function of t.

10.4.2 Write a program to read in $A(I)$, $B(I)$, where $I = 1, 2, 3, \ldots, 11$, values of $A(t)$ and $B(t)$ for $t = [0, .1, 1.]$:

$$\frac{dx}{dt} = e^{A(t)} \sin B(t), \quad x(0) = 1.$$

Tabulate t and corresponding values of $x(t)$.

In basic physics and calculus texts, velocity and acceleration are defined for a particle moving on a straight line (an x-axis) having position $x(t)$:

velocity: $$v(t) = \frac{dx}{dt}$$

$$\text{acceleration: } a(t) = \frac{dv}{dt}$$

If, for a particle moving on an x-axis, $a(t)$ is given, and initial values (at $t = t_0$) of velocity (v_0) and position (x_0) are known, then

$$\text{velocity: } v(t) = v_0 + \int_{t_0}^{t} a(z)\,dz$$

$$\text{position: } x(t) = v_0 + \int_{t_0}^{t} v(z)\,dz$$

Some classical examples follow:

1. For "the falling body,"

$$a(t) = -g, \text{ a constant}, \quad v(0) = v_0, \quad x(0) = x_0$$

Then

$$v(t) = v_0 - gt$$
$$x(t) = x_0 + v_0 t - \tfrac{1}{2}gt^2$$

2. For an "harmonic oscillator,"

$$a(t) = -\sin t, \quad v(0) = 1, \quad x(0) = 0$$

Then

$$v(t) = \cos t$$
$$x(t) = \sin t = -a(t)$$

Using the algorithms of numerical integration, we can extend our scope of assignments for acceleration and at least compute approximate solutions for velocity and position.

10.4.3 An observer moves on an x-axis with velocity $\sqrt{\sin t}$ and an object moves on the same line with velocity $\sqrt{2 + \cos t}$. If both are at $x = 0$ when $t = 0$, tabulate the distance between them for $t = [0, .1, 3.]$.

10.4.4 A u-axis and a w-axis intersect at angle θ. A particle moves on each axis starting at the vertex when $t = 0$. Write a program to read in θ and values of the velocities $V_u(t)$ and $V_w(t)$ for $t = [0., .1, 1.]$, and print out the distance between the particles $z(t)$ for each of these 11 values of t.

10.4.5 A particle moves to the right on an x-axis with acceleration $a(t) = e^{-t^2}$, and starts from rest at the origin at $t = 0$. Tabulate for $t = [0, .1, 1.]$ $a(t)$, $v(t)$, and $x(t)$.

10.4.6 A particle moves on the graph of $y = x^2$ with horizontal component of its velocity $v_x(t)$. Write a program to read in values of $v_x(t)$ for $t = [0, .1, 1.]$ and let $x(0) = 0$. Print out a tabulation of

$$t, \quad x(t), \quad y(t), \quad v_x(t), \quad v_y(t), \quad ds/dt$$

where $ds/dt = \sqrt{v_x^2 + v_y^2}$ is the speed of the particle.

10.4.7 A particle moves on the coordinate plane such that

$$\frac{d^2x}{dt^2} = a_x(t), \quad \frac{d^2y}{dt^2} = a_y(t)$$

$$v_x(0) = v_{x0}, \quad v_y(0) = v_{y0}$$

$$x(0) = x_0, \quad y(0) = y_0$$

Read in v_{x0}, v_{y0}, x_0, y_0, H, T and $a_x(t)$, $a_y(t)$ for $t = [0, H, T]$. Print out:

$$t: \quad a(t) = \sqrt{a_x^2(t) + a_y^2(t)}, \quad \theta(t) = \tan^{-1} \frac{a_y(t)}{a_x(t)}$$

$$v(t) = \sqrt{v_x^2(t) + v_y^2(t)}, \quad \phi(t) = \tan^{-1} \frac{v_y(t)}{v_x(t)}$$

$$x(t)$$

$$y(t)$$

A law of mechanics relates the mass of an object $m(t)$, its velocity $v(t)$, and a force $f(t)$ applied to the object:

$$\frac{d}{dt} m(t)v(t) = f(t)$$

The product mv is called the *momentum of the object.*

Consider the following special case: If m is constant,

$$f(t) = m\frac{dv}{dt} = ma(t)$$

For example, if $m = m_0$ and $v(0) = x(0) = 0$ and $f(t) = f_0 \cos t$,

$$a(t) = \frac{f_0}{m_0} \cos t$$

$$v(t) = \frac{f_0}{m_0} \sin t$$

$$x(t) = \frac{f_0}{m_0}(1 - \cos t)$$

It is interesting to sketch the graphs of $a(t)$, $v(t)$, and $x(t)$ on the same axes and observe the relationships between these functions.

10.4.8 A force acting on a body is

$$f(t) = \frac{1}{t \sin t}$$

and its velocity (and momentum) is zero when $t = 1$. Tabulate mv for $t = [1., .1, 2.]$.

10.4.9 A rocket loses mass according to $m(t) = 1 + 5e^{-2t}$ and the propelling force varies as $f(t) = -1 + 10te^{-t}$. Find the value t_1 for which $f(t)$ becomes positive. Tabulate $v(t)$ and $x(t)$ for $t = [t_1, .1, 2.]$

10.4.10 Tabulate $v(t)$ and $x(t)$ for $t = [0, .1, 3.]$, if $v(0) = x(0) = 0$ and

$$m(t) = \begin{cases} 2, & 0 \leqq t \leqq 1. \\ 1, & t > 1. \end{cases}, \qquad f(t) = \frac{\sqrt{t}}{t^2 + 1}$$

If the initial-value problem is described as: Find $y = g(x)$ if

$$\frac{dy}{dx} = f(y), \quad y(x_0) = y_0$$

the equivalent integral form is

$$G(y) = x_0 + \int_{y_0}^{y} \frac{dz}{f(z)} = x$$

where y expressed explicity in terms of x is

$$y = g(x)$$

The tabulation of the integral (usually for equally spaced values of y) is a tabulation of the inverse function $x = G(y)$:

$y = g(x)$	$G(y) = x$
y_0	x_0
.	.
.	.
.	.

To determine values of y for equally spaced values of the independent variable x, it is necessary to find an approximating function for $y(x)$ (usually a polynomial) and interpolate y for equally spaced values of x.

10.4.11 A model for population growth is

$$\frac{dx}{dt} = kx^{\alpha}(m - x)^{\beta}$$

where x is the population at time t. k, α, β, and m are constants depending upon the organism and its environment.

For $k = .05$, $m = 300 * 10^6$, $\alpha = 1.10$, $\beta = 0.80$, and the initial conditions $t_0 = 0$, $x_0 = 60 * 10^6$, tabulate t for $x = x_0 + k * 20 * 10^6$, where $k = 0, 1, 2, \ldots, 8$. Estimate the time when the population will be $150 * 10^6$. When $t = 50$, what is the approximate population?

10.4.12 A chemical reaction follows:

$$a(\text{units of } u) + b(\text{units of } v) \longrightarrow a + b(\text{units of } w)$$

Let $x(t)$ be the units of mass of w at time t. If $x(0) = 0$ and

$$\frac{dx}{dt} = k\left(u_0 - \frac{a}{a+b}x\right)\left(v_0 - \frac{b}{a+b}x\right)$$

where k is a constant pending on the properties of ingredients u and v and how they are mixed, u_0 and v_0 are initial units of mass of substances u and v.

If $k = 2$, $u_0 = 1$, $v_0 = 1$, and $a = 2$, $b = 3$, tabulate t for $x = [0., .1, 1.5]$.

Can this question be answered without numerical integration; that is, can t be expressed in terms of a formula of elementary functions and then x solved explicitly in terms of t?

REFERENCES

Computing Machinery

CHAPIN, N., *An Introduction to Automatic Computers*, New York, Van Nostrand Reinhold Company, 1963.

International Business Machines, *Introduction to Data Processing Systems*, Publ. No. C20-16840-2, White Plains, N.Y., 1969.

MAISEL, HERBERT, *Introduction to Electronic Digital Computers*, Chaps. 1–4, New York, McGraw-Hill Book Company, 1969.

RODGERS, W. H., *Think, A Biography of the Watsons and I.B.M.*, New York, Stein & Day Publishers, 1969.

ROSEN, SAUL, Electronic Conputers, A Historical Survey, *Computing Surveys*, (*Assoc. Compt. Mach.*), Vol. 1, No. 1, 1969.

RUSCH, R. B., *Computers: Their History and How They Work*, New York, Simon and Schuster, 1969.

SACKMAN, HAROLD, *Computers, Systems Science, and Evolving Society*, New York, John Wiley & Sons, Inc., 1967.

Programming

ANDERSON, DECIMA, *Computer Programming, Fortran IV*, New York, Appleton-Century-Crofts, 1966.

HOLDEN, H. L., *Introduction to Fortran IV*, New York, The Macmillan Company, 1970.

MAISEL, HERBERT, *Introduction to Electronic Digital Computers*, Part 4, New York McGraw-Hill Book Company, 1969.

MCCRACKEN, D. D., *Guide to Fortran IV Programming*, New York, John Wiley & Sons, Inc., 1965.

MURRILL, P. W. and SMITH, C. L., *Fortran IV Programming for Engineers and Scientists*, Scranton, Pa., International Textbook Company—College Division, 1969.

ROSEN, SAUL, *Programming Systems and Languages:* Paper #1, A Historical Survey; Paper #2, Some Recent Developments, New York, McGraw-Hill Book Company, 1967.

Programming and Problems

ANDREE, R. V., *Computer Programming and Related Mathematics*, New York, John Wiley & Sons, Inc., 1967.

BECKETT, BOYCE, and HUNT, JAMES, *Numerical Calculations and Algorithms*, New York, McGraw-Hill Book Company, 1967.

CRESS, PAUL, DIRKSEN, PAUL, and GRAHAM, J. W., *Fortran IV with Watfor and Watfiv*, Englewood Cliffs, N. J., Prentice-Hall, Inc., 1970.

DORN, W. S., and GREENBERG, H. J., *Mathematics and Computing*, New York, John Wiley & Sons, Inc., 1967.

GOLDEN, J. T., *Fortran IV Programming and Computing*, Englewood Cliffs, N.J., Prentice-Hall, Inc., 1965.

MCCRACKEN, D. D., and DORN, W. S., *Numerical Methods and Fortran Programming*, New York, John Wiley & Sons, Inc., 1964.

SOUTHWORTH, R. W., and DELEEUW, S. L., *Digital Computation and Numerical Methods*, New York, McGraw-Hill Book Company, 1965.

WILF, H. S., *Programming for the Digital Computer in Fortran*, Reading, Mass., Addison-Wesley Publishing Company, Inc., 1969.

Problems

GRUENBERGER, FRED, and JAFFRAY, GEORGE, *Problems for Computer Solution*, New York, John Wiley & Sons, Inc., 1965.

Computer Science

STERLING, T. D., and POLLACK, S. V., *Computing and Computer Science*, New York The Macmillan Company, 1970.

WEGNER, PETER, *Programming Languages, Information Structures, and Machine Organization*, New York, McGraw-Hill Book Company, 1968.

Computers and the Calculus Course

HAMMING, R. W., *Calculus and the Computer Revolution*, Boston, Houghton Mifflin Company, 1968.

STENBERG, W. B., and WALKER, W. S., *Calculus, A Computer Oriented Presentation, Center for Research in College Instruction of Science and Mathematics (CRICISAM)*, Florida State College, Tallahassee, Fla., 1970.

Numerical Analysis

CONTE, S. D., *Elementary Numerical Analysis*, New York, McGraw-Hill Book Company, 1965.

HAMMING, R. W., *Numerical Methods for Scientists and Engineers*, New York, McGraw-Hill Book Company, 1962.

HENRICI, PETER, *Elements of Numerical Analysis*, New York, John Wiley & Sons, Inc., 1964.

KUNZ, K. S., *Numerical Analysis*, New York, McGraw-Hill Book Company, 1957.

LANCZOS, CORNELIUS, *Applied Analysis*, Englewood Cliffs, N.J., Prentice-Hall, Inc., 1956.

MILNE, W. E., *Numerical Calculus*, Princeton, N.J., Princeton University Press, 1949.

MOURSUND, D. G., and DURIS, C. S., *Elementary Theory and Application of Numerical Analysis*, New York, McGraw-Hill Book Company, 1967.

Tables

ABRAMOWITZ, MILTON, and STEGUN, I. A., eds., *Handbook of Mathematical Functions*, New York Dover Publications, Inc., 1965.

Periodicals Oriented Toward Computation

A.C.M. (Association for Computing Machinery)

Journal (Research on numerical analysis and computer science).

Communications (Current events, algorithms, expository articles of general interest).

Computing Reviews (Of research literature).

Computing Surveys (Expository articles).

SIGCSE Bulletin (Special interest group on computer science education).

S.I.A.M (Society for Industrial and Applied Mathematics)

Journal on Numerical Analysis (Research).

Review (Expository articles, current events).

A.M.S. (American Mathematical Society)

Mathematics of Computation.

N.B.S (National Bureau of Standards)

Journal of Research—B (*Mathematics and Physics*).

Datamation (A commercial, bi-monthly magazine containing expository and news articles of interest of those in scientific analysis and data processing).

INDEX